BEHIND THE BITS

Managing The Media Maze

Frank M. Oppenheim
Helen C. Swift

University Press of America,® Inc.
Lanham • New York • Oxford

University Press of America,® Inc.
4720 Boston Way
Lanham, Maryland 20706

12 Hid's Copse Rd.
Cummor Hill, Oxford OX2 9JJ

Printed in the United States of America
British Library Cataloging in Publication Information Available

Library of Congress Cataloging-in-Publication Data

Oppenheim, Frank M.
Behind the bits : managing the media maze / Frank M. Oppenheim.
Helen C. Swift.
p. cm.
Includes bibliographical references and index.
l. Mass media. I. Swift, Helen. II. Title.
P90.066 1998 302.23—dc21 98-22384 CIP

ISBN 0-7618-1170-2 (pbk: alk. ppr.)

™ The paper used in this publication meet the minimum requirements of American National Standard for information Sciences—Permanence of Paper for Printed Library Materials, ANSI Z39.48—1984

CONTENTS

INTRODUCTION

If you browse among the shelves in the media section of your library, you will find a vast array of books--some focused on media techniques and skills, some on violence and pornography in the media, and still others on audience reaction to various aspects of the media. Some books bash the media finding nothing good to say about any of its many forms. Other books avoid anything negative and praise all the good qualities.

We take a different approach. Rather than focusing on audience reaction, we start with the traditional view that media of all kinds--TV, newspapers, cassette recordings, radio, books, movies, releases from public relations departments, and most recently, Internet, CD-Roms, and interactive `virtual' experiences--all present us with bits and fragments of data which make up the information packages. Billions and billions of such packages create the media maze. We ordinary citizens are called to negotiate our way through that maze and manage it as best we can. We are also called to hold publicly accountable those who manage the creation of this media maze.

We might consider the stimulus provided by bits of information and the response evoked in the receiver of the stimulus. We believe, however, that audiences need an awareness that goes beyond stimulus-response. We have grown so accustomed to being bombarded with information from all sides that we rarely, if ever, question the choice of these bits. Why do we receive these particular bytes at this moment rather than other, almost limitless, possibilities? Who determines what we see on television, read in the newspaper or hear on the radio? What are the motives for the choices these invisible individuals make? What accountability do they feel toward their audience? These questions form another level of inquiry, one we attempt to focus on in this book.

We believe in the need to avoid simply surrendering to the media and to those who create the media. Our freedom demands an alertness to the subtle influence of the media in our lives. This means examining critically the attitudes and values presented by the media as well as the strategies and techniques of persuasion found there. We also need to examine ways of participating in creating media programs and in ways of calling media leaders periodically to public accountability.

Different accounts of the media arise from the various perspectives held by different authors. If a liberationist or Marxist or feminist perspective governed the present work, the reader would hold a far different book in his or her hands. This book, however, arises from the perspective of Christian humanists. So we maintain faith in natural creation, especially in human persons who, despite their shortcomings, have tremendous potentials and value. We also believe in the basic goodness of the things that humans create--like TV, computer programs, and the World Wide Web. With this perspective the present authors invite readers to enter into our belief that eventually most persons <u>tend to</u> adopt wiser, healthier ways and have a deep affirmative commitment towards all persons involved in the media world. This commitment extends to media owners, directors and others called to serve the human development of their audiences: children, students, and adults--both the middle aged and the more elderly.

A brief overview of the media in Chapter 1 helps to situate the present media reality. Beginning with print in the 17th century , we take a summary view of the way ever newer and more sophisticated developments in technology have challenged existing media. Radio changed the way news was presented in newspapers. TV newscasts and programming have had an impact on the importance of radio. In the world of entertainment TV and movies vie for audience time. More recently, Cable TV has threatened the networks by providing a greater variety of programming. The trend to smaller and more user-friendly devices attracts all of us in varying degrees. Computers giving access to the "information highway" are already creating a gap between the haves and the have-nots. Those who can use a computer earn 15% more than others. In order to be in control of this fascination rather than let it become an obsession and control us, we need an awareness of some of the physical and psychological effects of the media, especially of TV, on the individual.

Chapter 2 treats some of the positive ways TV influences our lives and our culture. There is a tendency to concentrate on the negative aspects of TV--violence, sex, and the pull to greater consumerism-- while overlooking the good aspects of this technology. Both radio and TV have shifted communication away from the literate mode to greater orality. This shift affects the way we relate to people in daily life and can prevent many misunderstandings. Television has changed the way politicians appeal to their audiences and the president's ability to communicate to the nation and the world.

In Chapter 3 we consider how advertising reflects the values of those behind the scenes. Advertisers rely on carefully devised psychological and sociological studies to reveal the buying habits of various groups of Americans. Armed with this information, they can construct commercials that appeal to a particular group of consumers. For instance, a change is taking place in marketing as millions of computer users are attracting the attention of advertisers. Thousands of companies are eager to reach the audience of upwardly mobile consumers (average income: $55,000) through online advertising. We notice specific pressures--some subtle and some not so subtle--used by the media to get Americans to buy ever more.

Each day more notable events happen throughout the world than the evening news can possibly report. Why, then, are we presented with this particular event rather than one of the countless other possibilities? Besides many other factors discussed in Chapters 4 and 5, Ed Joyce, former president of CBS News, admits that the competition for high ratings and the personalities of anchormen shape what the viewer receives as news.[1] This attitude tends to dramatize the news and make it entertaining. In this way, sponsors are not lost and revenues do not slump. The "factual world," however, is thus replaced by a world constructed for entertainment and profit.

In our age of super-communication information, increases exponentially. Paradoxically, however, the American public finds so much information unavailable to it. How explain this blockage? The 6.8 million documents made unavailable each year by the executive branch of the government is only the tip of the iceberg. Add to this number the secret tests conducted on civilians by the military, the secret activities of the CIA, and documents of other agencies and the number becomes astronomical. The only information becoming more available is data on the private lives of citizens. The control of information which we can access and the increasing information about private citizens available through the intrusion of the media are highlighted in Chapter 6.

Chapter 7 explores the roles of government, business, and the hierarchy of the media. The belief that top level executives in the TV, newspaper, radio, book, and movie industries are free in their decision-making is simplistic. As we shall see the reality is much more complicated. Constraints involving many governmental agencies, big businesses, and competition within the media industry itself are important factors influencing the product presented to the public.

The situation explored in Chapter 7, helps us to realize more clearly the impact of the media on its audience, the subject of Chapter 8. Here we are focusing primarily on TV, movies, radio, and newspapers as the forms of media most pervasive in the lives of ordinary citizens. The media affect in a subtly powerful, but usually undetected way what people think, the values they uphold, and the lifestyle they desire. The strong reaction of many people against the excessive violence on TV and in the movies tends to obscure other influences that may be just as destructive. By examining the subtle global instruments which the media use to control attitudes and choices, Chapter 9 completes our exploration of what is behind the media maze. The influence brought about by the ownership of large media systems is dwarfed by the power of dominant firms, their interlocks of directorates, and conglomerates. Through these methods of concentrating power a relatively few individuals control the information highway, not only in the U.S. but throughout the world. Taylor Caldwell's imaginative novel, *Captains and Kings,* has become a reality. As we move into the future, we are aware of the concerted efforts of the U.S. to turn computers into weapons of destruction, but less aware that scientists in other countries are also investigating this possibility.

Realizing that no one can accurately predict the future, we look ahead in Chapter 10 at some of the shifts in culture that we my expect as the media continue to invade every aspect of our lives. This chapter develops briefly the various dimensions of the transformation taking place today: linguistic, moral, philosophical, spatial, political, military, medical, economic, financial, commercial, educational, psychological, ecclesial, scientific, artistic, and religious. All these dimensions taken together make up what we experience as culture. If we do not want to be passive onlookers in our world of so many fascinating possibilities, we need both to know what is happening and to exert our influence.

Chapter 11 contains some suggestions for reaping the benefits of an intelligent use of the media. Beginning with education and establishing a new view of communication, individuals and groups working together can take responsibility for their future. As an aid in this endeavor, questions for discussion are included at the end of each chapter. Wise choices now can prevent costly mistakes in the future. No one person has the wisdom and knowledge necessary to lead the public in wise choices, but many intelligent and alert individuals acting together can create the attitudes and value system for using the media's powerful tools to the best advantage. In this way the many dimensions of cultural transformation can be focused on providing a peaceful, fulfilled life for all, bringing the peoples of the world closer together in peace.

We are grateful to our friends and the media experts who have contributed to this book by their helpful suggestions and comments.

Frank M. Oppenheim
Helen C. Swift

Chapter 1

OVERVIEW OF THE MEDIA TODAY

Each evening, Monday through Friday, *Jeopardy* fans crowd around their TVs to watch three competitors respond to categories of trivia. Besides a broad general knowledge, success depends on two abilities: instant recall of isolated facts and a physical stimulus-response system that reacts faster than normal. In many ways, this game show symbolizes American culture today. Disparate bits of information inundate us and leave no time for reflection on meanings. We live reactively, spontaneously responding to the latest stimulus, rarely making a deliberate reflective response.

Early Developments of Media in the U.S.

A brief look at the way media developed in our pluralistic American culture contributes to a better understanding of the place of mass media in our lives today. It also lays the foundation for responsible choices as media technology becomes more pervasive and complex.

Although in the West, the printing press was invented in 1451, it was not until 1639 that the first press began operations in the colonies of the New World. Thus 1639 is an historical date as it records the very first step in North America toward modern media technology. Apparently, printing was not top priority among these early settlers as it took more than 50 years before the first newspaper appeared. It was entitled *"Public Occurrences Both Foreign and Domestick."*

By 1719, when Samuel and John Adams published the *"Boston Gazette,"* news was becoming more available to the ordinary citizen. No daily newspaper existed in the U.S., however, until 1783 when the

"*Pennsylvania Evening Post*" reached the streets. For at least another 50 years until Samuel Morse developed the telegraph, only local news could be considered as at all up-to-date.[1] As we shall see in Chapter 6, the telegraph not only made news from afar quickly available, but started shifting from linear thought patterns to a staccato mode and emphasized the power of having news from afar.

Media and Cultural Shifts

During the nineteenth century, colonization and immigration continued to characterize the American scene. Frequently, new immigrants were embarrassed and uneasy about their Old World origins. They often changed their names, learned English, and struggled to provide an Americanizing education for their children. The public school system expanded to meet the growing demand for universal compulsory education. At the same time, labor unions formed to protect the rights of workers. As immigrants made every effort to become integrated into American society, daily newspapers and general interest magazines helped to diminish the ethnicity and differences of people.[2] Many newspapers, however, catered to particular ethnic heritages; e.g., the Jewish press.

The Saturday Evening Post, founded in 1821, exemplified the quality of general interest magazines published during the middle 1800's. Its list of contributors included William Cullen Bryant, Harriet Beecher Stowe, James Fenimore Cooper, Ralph Waldo Emerson, Nathaniel Hawthorne, and Edgar Allen Poe--writers now studied in American Literature courses. As George Steiner has remarked, "The period from the French Revolution to World War I marked an oasis of quality in which great literature reached a mass audience."[3]

The years after 1850 ushered in the "technocratic" era. Alfred North Whitehead summed up the rapid growth of inventions after 1850 when he remarked that the idea of invention itself constituted the greatest invention of the nineteenth century.[4] Among the inventions of the second half of the nineteenth century that affected the media are the typewriter, the telephone switchboard, the radio, and motion pictures. Early in the twentieth century, sound was added to motion pictures and they became one of the most popular forms of entertainment. Little did the movie-goers realize as they watched the movie screen that they were also witnessing a shift in their culture.[5]

The so-called "technocratic" culture, a culture greatly influenced by technology, eroded the line between the intellectual interests of educated people and those of the working class.[6] Integration and homogeneity so characterized early twentieth century America that social historians refer to it as a melting pot.

During World War II, new technologies were developed for military use. Soon after that War ended, peace time applications appeared that radically changed American society. The availability of radio, and later TV, challenged newspapers to provide up-to-the-minute news. In order to stay alive, newspapers relied more on wire services, changed their format, and used color photography. Today, despite the availability of news on radio and TV, newspapers still have a place in most American homes. James Reston, a well-known journalist, attributes the survival of newspapers to the fact that big-city dailies generally are better written, better edited, and more accurate than in the 1940's when he became a cub reporter.[7] Also, technology has been unable to match newspapers availability to city readers who can scan the newspaper, read it during breakfast, or fold it up and carry it around to be read later with ease.[8]

After World War II, the gathering, filtering, and delivering of news were activities that only mass organizations could do. Every American received the same kind of news, whether in newspapers, on the radio, or TV. Today, however, expanded technology increasingly reverses the trend toward homogeneity. New methods of communication are causing a shift toward special interests. Now, newspapers add human interest stories and syndicated columns on specific subjects, such as health, cooking, and humor. They also add puzzles which attract some subscribers and expand local ads and the adventures of locally prominent stocks which appeal to local readers and thus increase revenue.

As TV began competing with radio, radio turned to talk shows and music, even specializing in the type of its music--country, Gospel, hard or soft rock. Television sets take the place of radios in living rooms, family rooms, dens, and even bedrooms, but radios still find a space in work areas, such as the kitchen, laundry, and shop.

Now the trend toward special interests is making its way into TV. Jim Rocca, partner of newly formed Entrepreneur Television, Inc.,a company that produces informative niche programming for TV, claims, "The age of mass audience TV is coming to an end and niche marketing will dominate the future." MIT's Media Lab is developing "The Daily Me," a system for programming computers to recognize news content and provide only stories that interest the individual user.[9]

Now with newspapers and radio, commercial TV also feels the pressure of competition. Cable has added many more choices for the ordinary viewer and soon interactive TV through the use of a computer will open up almost unlimited choices to the TV viewer. Other factors such as de-regulation and mergers, along with widened rights of syndication, enter into the future status of TV. Moreover, powerful forces in advertising and in Hollywood studios work hard to prevent the commercial networks from falling prey to pay-per-view channels, cable movie channels, videos, and public broadcasting. These alternate TV modes threaten advertising since they do not contain commercials.

The public, however, needs the commercial networks. Despite their many failings, these networks occasionally create a sense of community that resonates with some of the togetherness experienced in the late 1800's and early 1900's. For example, John Kennedy's funeral and Martin Luther King, Jr.'s assassination brought people across the nation together in shared experiences of grief. Likewise, the miniseries, *Roots*, provided an educational experience shared by three fourths of all viewers. More than a hundred million TV viewers thrilled as the astronauts in outer space broadcast the picture of our planet as a blue, sunlit sphere floating in the darkness of galactic space. Television united those viewers in knowing and feeling how small and frail is the planet on which we must live together. More recently, people around the world shared the wonder and joy at the fall of the Berlin Wall. As Ev Dennis, executive director of Gannett Center for Media Studies at Columbia University, observed: "We have shared values that are enhanced by three networks. . . . It [TV] does create a national consensus of values and of what we think is important."[10]

In the future, if TV can show space ships from various nations moving peacefully to and from different space stations, exchanging passengers and goods, and following international law, then our sense of global community will be awakened. Still further in the future, and despite various resistences and terrorism, the networks can even gradually promote an international consensus. They can also foster those central attitudes and basic values which all peoples regard important for sharing life together.

Television and Daily Life

Despite those rare specific events that can unify Americans, the glut of information and its lack of integration increasingly fragment American opinions. Social scientists do not agree on the significance of this trend or its long range consequences. In the multiplication of special interests, some social scientists see only selfishness; others note more positive results, especially in the field of education.

In general, education benefits from developments in communication technology which greatly expand the educational field. These new technologies let students make the most of some of their talents and skills and call for balanced growth in other humanizing potentials. Society at large benefits when such higher qualified students graduate from high school and when more go on to college and post-graduate work.

If individuals can pursue their own interests, they become more creative and inventive. Teachers of young children give them many opportunities in school to become aware of the vast possibilities before them in the world. Teachers and parents encourage children's hobbies, and their work on projects for science fairs. They show their appreciation by displaying the children's creative work.

Television can be a powerful educational tool both in schools and for the general public. For example, computer-trained students who have TV monitors and are online with the Internet hold the key to amazingly rich data banks and to almost infinite research possibilities. The use of TV in space has become so common that the sight of astronauts at work in their space capsule no longer startles us. We take for granted that TV cameras mounted on satellites show us the moon or distant planets. Television cameras also reach down into the deep waters of the oceans, revealing life forms never before viewed by scientists.

Throughout this book special emphasis will be given to TV as today's most influential media, although its dominant position is daily being challenged by the Internet and World Wide Web. In American society TV images give vivid concrete embodiment to some of the intangible beliefs of our culture. For example, much of the prime time programming presents drama and comedy against a background of luxury and wealth, re-enforcing the belief that more is better. These images thus provide a basis for a knowledge of our beliefs and motivation to implement those beliefs.[11]

Looking closely at different kinds of TV programs, Gregor Goethals discerns residual elements of traditional imperial and religious symbols of authority--mythic, informational, and political. She explains:

> First, American values are personified in heroes and heroines of sit-coms, dramas, soap operas, and commercials. Like ancient myths and medieval legends, these popular forms tell easily understood stories

> of love and hate, birth and death, success and failure, loyalty and betrayal. Second, nightly news programs map the cosmos with images that define for us the contours of reality. News commentators, like ancient priests, lead us through ceremonies that purport to represent and explain worldwide events. Third, coverage of our political leaders has created authoritative images comparable to the statuary of the ancient world; television's "photo opportunities" have taken the place of sculpted portraits of rulers like Pericles and Caesar Augustus.[12]

The general public also profits by the use of TV in commerce and the medical field. Tiny TV cameras ensure safety in banks, subways, and malls. Other minuscule TV cameras lead to a more accurate diagnosis than the exploratory surgery of just a few years ago. These uses of TV may be overlooked because many people tend to look upon TV primarily as entertainment.

Many authors have written, often erroneously, of the adverse influence of TV on viewers. Studies show, however, that the well developed human capacity for selective attention protects viewers from much of this adverse influence. Bombarded with 1,600 ads per day, the average individual responds--not necessarily positively--to only 12 of them. The mind filters signals and the brain responds consciously to only one sensory stimulus among each million stimuli being sent.[13]

One of the most detailed experimental field studies yet published reveals that under ideal conditions multiple exposures to an advertising campaign yields, at most, a 2% increase in sales. Moreover, more detailed studies of the link between advertising and behavior show how seldom increases in sales can be attributed to advertising exposure. Study after study reveals that advertising produces very small effects, or highly conditioned effects, or no effects. Why then bombard viewers with commercials? The myth of pervasive advertising survives because the critics of advertising fear doing battle with advertising executives linked to media, and even to the manufacturers themselves.[14] The media depend financially on advertising. As we shall see in Chapter 9, the links formed by mergers, conglomerates, and interlocking directorates are so involved and complex that changing the financial base of the media America's largest industry, would shake up the whole economy.

Paul Klein's model of TV viewing, "Least Objectionable Program," remains the classic caricature of our media behavior. He describes the viewer as seldom deciding on a particular program. Rather, for the great bulk of viewers, the process follows these steps. First, one decides to watch TV. Then, one turns on the set and flips through the channels to check the possibilities. Then, one watches whatever one finds least objectionable. Even if nothing attracts the viewer, one rarely turns off the set. Klein's speculations turn out to be supported by a great deal of research and apparently characterize people's behavior with other media as well.[15] Studies show that for at least one third of the time, TV viewing provides a background noise while individuals talk, read, answer the telephone, eat, or take care of household chores.[16]

Development of Interactivity of the Media

Just as newspapers and radio had to make room for TV, new technologies now challenge TV to adapt or be pushed aside. New technologies place emphasis on interactivity in which both sender and receiver share control over the communication process.[17] The power and flexibility of interactivity offer unprecedented potentials in the field of education. Computer-based media allow the student to manipulate themes creatively and to experience how basic principles work themselves out in practice. In these ways the learning process is made both more effective and more enjoyable. Also, the use of computer-based learning environments allows the student to take greater control of the pace and style of learning.

Computers can present "micro-worlds" allowing the students to respond to various simulated situations. Observing the consequences of their responses, students can learn patterns of wise choices and long-range planning. The computer enhances the value of traditional role playing by simulating a variety of "What if. . . ?" situations calling for specific responses. Used in this way, computers enrich and re-enforce the learning of many concepts and behavioral patterns.

With the use of the computer, students have access to programmed concepts and have the ability to manipulate these concepts. Micro-world learning features the ability to transfer a concept from one representational form to another, adding greater variety and re-enforcement to the learning process. It has proven especially effective in preventing substance abuse.[18]

Computer-based learning, however, leaves some basic abilities uncultivated. It does not instill those values so important to humanized living: insisting on fair treatment of others rather than depriving others by seizing all advantages for oneself or being kind, gentle, and compassionate to everyone, especially to the poor and suffering. Neither does it instill the prudence and discretion necessary for wise decisions.

Moreover, if the software is less than ideal, it adds frustration to the learning process. Then too, some teachers and students feel threatened or confused or hindered by the computer. In brief, interactive media provides a power tool but not an antidote for the difficulties of the educational process. As even the computer programmer realizes, the computer is never a replacement for the thoughtful and talented teacher.[19]

Electronic media may also help us learn our civic responsibilities. In 1987, Christopher Arterton examined twenty demonstration projects that used new communications technologies to facilitate public participation in policy-making. In some projects, members of the U.S. Congress participated in computer conferences with their constituents. The enthusiasm and energy, as well as the technical resources, invested in those diverse projects impressed Arterton.

Another type of civic responsibility project brought community groups together by two-way cable. Satellite technology connected remote villages; electronic town meetings included debates, polls, and plebiscites. In almost every case, the enthusiastic initiator of these projects used the new technological means for an idealistically democratic end. Sometimes, with great effort, they succeeded in attracting public attention and stimulating thought, debate, and participation. At other times, the considerable talents and resources of the organizers met apathy and inertia, leaving the political process largely unaffected. In every case but one, at the completion of the demonstration project, further action halted. The organizers wrote their reports and moved on to other projects.[20]

The one activity that did continue under its own power involved a mayor and community groups who continued to use community cable TV effectively for public discussion. The mayor believed that political benefits flowed from the activity and she felt motivated to continue it.[21]

Those who made the effort to use community cable TV effectively might, at a first hearing, be discouraged by the too easy but insightful prophesy of Andrew Lippman of Media Lab:

> Forget television sets. In three years there won't be any. Instead they will be computers with high-quality display screens. Inside these computers there will be digital instructions allowing them to receive ABC, NBC, BBC and anything we can dream of.[22]

Six Goals for New Electronic Technology

In his book *Powershift,* Alvin Toffler describes six features that the new technology needs to integrate.[23] He says that to remain competitive the new electronic technology must be interactive, small enough to be portable, mode-conversive, interconnective, ubiquitous, and global in impact. He clarifies these six traits as follows.

Electronic equipment reaches acceptable interactivity when it allows broadcaster and receivers together to control the creation of a media experience. For example, a TV broadcast system becomes genuinely interactive when both the network with its local station and the receiving audience with their computers and phones mutually contribute direction and input to create a program.

The demand for compact mobility has created small equipment like laptop computers and notebook word-processors. It has led to portable X-ray machines and cordless phones that allow people to communicate from almost anywhere to almost anywhere. Joined to this demand to be simpler and handier is the demand to become less expensive and more marketable. The results are fax machines, for cars, vest-pocket videos, portable printers and copiers. Smaller and smaller micro-chips, carrying vaster amounts of information, make possible all this compact mobile equipment.

Besides being interactive and mobile, the new technology seeks to be mode-conversive, that is, facilitating transfers of information from one medium to another. Speech information can already be converted into print and many hope that automatic translation from one language to another will soon be readily available.

Interconnective technology overcomes trade barriers to the flow of information. At present, devices from different manufacturers cannot always be connected with those of another manufacturer. In the future, interconnectivity will provide for devices being connected regardless of manufacturer or country of origin. One basis for such interconnectivity is a common system of measurement. The British have adopted the metric system to trade better in the European market. This became mandatory, in general, as of October 1, 1995, excepting the pound for grain and vegetables until 1999, and with an indefinite exception of the pint for beer and miles for road signs.[24] Americans can avoid a certain self-isolation in global trading by adopting the metric system and thus promoting the interconnectivity of products.

Of particular concern to political scientists is the ubiquitization of technology, that is, the systematic spread of the new media system around the world and down through every economic layer of society. Alvin Toffler warns of the threat to governments unless this worldwide spread takes place. He writes:

> A potential nightmare facing high-tech governments derives from the split-up of populations into the info-rich and the info-poor. Any government that fails to take concrete action to avoid this division courts political upheaval in the future. Yet this dangerous polarization is hardly inevitable. . . . It is, therefore, in the distinct self-interest of the affluent to find ways of extending the new systems to include, rather than exclude, the less affluent.[25]

Since electronic media have internationalized business, culture, and education, a final feature of the new technology is its global character. Just as capital flows electronically across national borders, so, as information also becomes internationalized, it will follow complex pathways to all parts of the globe via satellites, fax, and electronic mail.

Toffler sums up the effect of these features of the new media by saying:

> The combination of these six principles produces a revolutionary nervous system for the planet, capable of handling vastly enlarged quantities of data, information, and knowledge at much faster transmission and processing rates. It is a far more adaptable, intelligent, and complex nervous system for the human race than ever before imagined.[26]

With this new electronic nervous system we will move into the future. Clearly electronic technology can become a rich blessing for humankind or its horrible nightmare or a mixture of both. Because of a lack of awareness and intelligent choices, electronic technology--particularly in its TV form--has been until the present a mixed blessing, as the following chapters will show.

Moving Ahead

Critics of Toffler sometimes accuse him of painting an inaccurate picture of the future. However, he seems to be on target when he and other sociologists envision a powerful upsurge in knowledge through the new speedier technology. Toffler writes:

> [W]e are witnessing one of the most important changes in the history of power. For it is now indisputable that knowledge, the source of the highest-quality of power of all, is gaining importance with every fleeting nanosecond [that is, one billionth part of a second].[27]

As we move into the future of a more rapid spread of information throughout the world by new technology, the positive possibilities for educational and cultural enhancement stagger the imagination. The Renaissance and the Enlightenment challenged the human mind but our Electronic Age with its powerful tools challenges our powers of perception and our ability to choose wisely in managing our media maze. We may not allow just a few individuals to meet and deal with this challenge. A critical mass of many alert individuals needs to arise whose awareness of past mistakes and of wiser choices enables them to participate in the democratic process of controlling the new technology. By looking behind the bits of information that make up our daily fare, the authors attempt to throw some light on our future choices so that we may choose wisely and avoid basic mistakes in managing the media maze.

Before looking at some specific ways in which media influence our lives, we will turn first to TV, the most influential of media, and consider the evidence for deliberately preferring a positive attitude towards it.

QUESTIONS FOR DISCUSSION

1. How have you experienced a sense of community--familial or national or global-- created by the various TV networks (traditional: ABC, CBS, NBC or the more recent, Fox, WWW, CNN, etc.)?

2. How do you respond to the daily glut of information?

3. Studies seem to indicate that TV commercials have very little effect on sales of a product. Do you agree? Why or why not?

4. What value do you place on the use, in the classroom, of electronic equipment--computers, TV monitors, movies, etc.? Does the aim of promoting learning or of entertaining generally dominate in such use? On what evidence do you base your response?

5. Gregor Goethals notices that TV programs convey symbols of mythic, informational, and political authority. Which examples of such symbols of authority do you find in the TV programs you watch?

6. According to Alvin Toffler, the new technology needs to integrate six features. Of these six, which seems the most urgently needed to you? Why?

7. Toffler claims that knowledge is the source of the highest quality of power. In what sense is this true?

Chapter 2

CHOICE OF A PRIMARILY POSITIVE ATTITUDE TOWARDS TELEVISION

A famous rabbi pointed out, "You notice the splinter in your neighbor's eye but do not perceive the wooden beam in your own eye."[1] When applied to life today, this warning cautions users of the media, and especially viewers of TV, against slipping too quickly into a primarily negative view of the media. As proved by the thrust of many newscasts and newspaper articles, Americans are being schooled to find fault with certain institutions, to blame other people, and even to judge them guilty, without much search for "the wooden beam in one's own eye." We know, moreover, that TV is currently the most widespread and influential of media. Accordingly, an urgent need arises in this introductory section to our study of the media to examine critically which basic attitude we need to adopt toward TV and why. For one's basic attitude toward TV acts like colored eye-glasses that insert a tinted filter--light or dark--into everything one views in the world of media

Today many find it hard to imagine life before TV. It has become so commonplace and taken for granted. Some people, puzzled by this change, ask, "Back in pre-TV days how did mothers keep their little children and also do their housework? What did the elderly in retirement and nursing homes do all day? How did sick people in hospitals survive their stay there without TV to watch?"

Do ordinary people, however, really ask themselves how this atmosphere of TV technology affects them? Of course, many media experts, psychologists, social scientists, and philosophers have examined how Americans interact with it.[2] For instance, the American Psychological

Association carried out a five-year study of TV's impact on certain groups of Americans.[3] Joseph R. Dominick has kept in touch with changes in the media through a series of five editions of his *The Dynamics of Mass Communication*.[4] He has focused on the social implications of viewing TV and probed the ethical issues arising from TV and other media broadcasts. The findings of the studies by the A.P.A. and Dominick alert non-experts in the media to a threefold call: to recognize the cultural shift effected by the new media, to accept it, and to seize the opportunity it offers. Each of these calls invites a closer look.

The Cultural Shift to Orality

In the late twentieth century, the increasingly pervasive use of sound bites and sight bites by recordings, movies, and especially TV, has created a profound shift in culture. More and more, people's minds are being nudged out of a print-based culture into an increasingly oral and visual kind of culture.[5] Although books, journals, and newspapers will not become extinct, they will attract a decreasing percentage of Americans. They will continue to be useful and even indispensable in the fields of literature and history, in research, and especially in the professions; for example, in law and university teaching. Moreover, print advertising has stimulated a significant amount of talk about items for sale in the past and does so today. Yet the world of ordinary communications is today becoming ever more oral. It even needs to grow more so if an international family of nations is to be nurtured humanely by increasing oral interpersonal relations. Media-based sound bites and sight bites have played and are playing a key role in bringing about this vast cultural change.

Amid this startling cultural shift, then, the challenge confronting people today is whether more and more of them will adjust to their new situation. Or will they tend to prefer the familiar routines of literate culture? Making this adjustment will undoubtedly call people to undergo a deep psychic change. While granting the indispensable need for the literary mode in certain fields, if they focus on the advantages of the oral form of culture, they may reach this adjustment more easily.

What, then, are some of the advantages that a primarily oral culture offers to a people that traditionally are literate? First, to achieve communication a primarily oral culture, focuses directly on interpersonal interaction rather than on the art of using written words skillfully. In a primarily oral culture the priority lies on the persons who communicate

The words they employ are, within limits, relativized. By focusing more on the interacting persons, oral communication takes into account their attitudes and choices more than the literate form does. The latter tends to highlight words more than the people who use them.

Then, too, persons who communicate orally find the process spontaneous, fluid, and flexible. The oral flow of words, gestures, and affective pressures moves people to become more flexible in the communication process. This increased flexibility helps people adapt more readily to a steadily flowing stream of gestures, words, and emotional tones. By contrast, when people lack these accompanying signals and are tied to words fixed on printed pages, their thinking tends to become fixed and inflexible. So, the more people become familiar with oral communication's greater flexibility, the more flexible they become in their own communication processes.

Furthermore, since it is more spontaneous, oral communication approaches far nearer than print to that natural home-style communication employed from infancy upwards during people's process of growing up. For instance, songs or stories heard either on mother's lap or proclaimed at national or religious holidays are easily recalled. For they were some of the earliest sounds to strike one's ears. Later on, to enrich one's college education, a person may have collected "audits" by tuning in on the more interesting professors. On jury duty, one may have listened to witnesses during a court trial.

Oral communication conveys more riches than does the literate mode. It conveys the particular ambiance of changing feelings, of approvals or disapprovals, of decisions and commands communicated during interpersonal exchanges. For grasping such a particular ambiance, the literate mode requires the correct selection of words and the more refined techniques of style. In addition, the latter focuses on the <u>what</u> being communicated more than on the <u>who</u> doing the communicating. Often it loses that rich atmosphere of associative, conative, and emotive meanings that are indispensably bound to oral communication.

We assume that advanced civilizations cannot survive without carefully written laws, contracts, and policies. Yet since literacy inserts paper between its communicators, it remains indirect and less interpersonal than directly personal oral communication. Literacy requires written or printed words to function as intermediaries between those trying to communicate. Granted, any word--spoken or written--carries some ambiguity and works partly out of its author's control. Yet once a word becomes written or printed, it not only fixes a meaning--e.g., in a law or contract--but it becomes more detached from its author and grows still more distant from him or her. It becomes less controllable by the author and more vulnerable to diverse interpretations. For if a written communication had not escaped from its oral, more spontaneous, and intimate mode, an orally united pair of speaker and listener would have more simply and immediately corrected any emergent misinterpretation of a word or phrase. Oral communication thus seems to foster a self-corrective process more easily and promptly.

Acceptance of the Cultural Shift

As mentioned, the movies, recordings, and especially TV have produced a worldwide shift toward orality. How, then, if we take fifty years as a rough watershed between young and old, what responses to this shift become clear? Most people under fifty have absorbed this cultural shift fairly well, although in varying degrees. Unfortunately, when one tallies up the percentages of today's illiterates and semi-literates, it seems that many of these folk under fifty abandoned the literary mode only too readily. By contrast, some die-hards over fifty need to recognize that TV and its VCR companion have long since "led the dance" in the world of communications and will continue to lead it. Meanwhile, most of these folk over fifty still appreciate the literary mode even while they accept the televisionization of American culture and recognize the power, limits, and opportunities of TV.

In contrast to the purely secular and dehumanizing stories now broadcast only too frequently, some skillful religiously-rooted people, like Bill Moyers, realize that they can tell <u>other</u> stories on TV. The style of their dramatically narrated sacred stories contrasts sharply with that of most broadcasting evangelists. These new dramatic narrators of bible stories sometimes depend on the Holy Spirit to mediate to others what their own religious life has empowered them to realize in themselves. They can revere life in all its forms, be grateful, forgive and be reconciled, work with and sacrifice for others. In this way they enhance the orality of their post-printing-press culture with a more humanizing orality--a culture they know they can create and feel responsible for creating. They also enrich their less religiously-rooted fellows who have a right to expect something more from them.

This positive endorsement of the electronic media, however, needs schooling from our past and present experiences with the media. We need to be realistic and discern the media carefully if we are to use creatively the current and future developments of TV and other electronic media. True, at present we encounter only too frequently in the media the toxic lyrics of some hot recordings, TV movies of sex-and-violence, and some sexily thumping guitar artists. Recently Senator Robert Dole had grounds for warning the media moguls who orchestrate American TV, movies, and musical recordings. When Dole branded some of their programs as "nightmares of depravity," many media moguls, stung by his critique, bellowed dissent but far more Americans rightly agreed with him.[6]

As a result, since the older generations have experienced both oral and literate cultures, they bear a greater responsibility for assessing the media in a penetrating realistic way. Furthermore, their richer past experiences call them to a more disciplined use of the media. Both of these responsibilities hold truer still for those who have studied the media. For especially at risk here are the younger generations and that multitude of those over fifty who have not studied the media's impact upon themselves. At even greater risk are those less mobile audiences, such as the institutionalized, the elderly, ethnic minorities, children, and women who spend most of their time at home.[7] Not overlooked here, however, are the benefits that TV's better programs bring, for instance, to those confined in elderly homes which are otherwise often depressing. For these shut-ins often gain psychic uplift from TV shows that entertain or excite interest. Their lonely hours are offset by TV voices from remote human companions. They also feel closer to distant friends if they know that they and their friends are viewing and responding to the same program.

In general, however, if Americans spend an average of three to four hours daily before a TV set, TV has to be gradually transforming their feelings, attitudes, values, and beliefs. At times even actively alert and mobile adults succumb to become "coach potatoes" before the TV set--say on New Year's Day or Superbowl Sunday or when for several hours after an exhausting workday they collapse before the TV set. How much more pervasively and profoundly does TV influence those other Americans who encounter its many defects and built-in limitations with minds less alert and discerning than those of adults who meet the TV set with fresh minds?

Clearly, then, the electronic media invade many people's lives and shape them profoundly. Yet, as longtime media student Neil Postman says, "We speak about America as if television has merely been added to it and little else has changed. Americans watch television, but we have not yet reached the point where we watch ourselves watch it."[8] That is, people find it hard to keep a reflective focus on TV and its impact on them because its influence is so pervasive and subtle.[9]

When people do decide to watch themselves watching TV, they can only too easily become preoccupied with its many negative effects. This draws them into a defensive strategy of containment. They complain about TV and readily tell themselves that they must protect themselves and their children from the low taste and lower morality of most TV shows. So, raising the drawbridge, they try to retreat into their uncontaminated castle, as if today they could be both human and yet not affected by TV.

In the long run, however, such a strategy of complaint and containment will not work. For TV and its offspring already permeate American culture and will saturate it yet more in the future. Today's kind of TV may grow obsolete, thanks to upcoming inventions in electronic media. Yet people of the twenty -first century will be living in an even more intimate and vivid media world, one that will teem with its bites of sound and sight, and possibly even of touch.

People's mainly negative attitudes towards TV arise because to a large extent they do not recognize its built-in limits and thus come to expect too much from it. So, these unrealistic expectations will be our next focus. Later, we will marshall evidence for a primarily positive attitude towards TV.

Unrealistic Expectations of TV

Americans need to face up to TV's intrinsic limits and stop expecting it to do things it cannot achieve. How many have even taken careful measure of what TV cannot do or asked themselves whether they expect too much of TV? Some self-styled educators, for example, expect TV to teach youngsters to think as critically as Socrates taught his students to do. Some humanists expect TV plays to penetrate into human nature as profoundly as Shakespeare's best-known dramas. Some professors want the speakers on TV to distinguish sentences as sharply as did Wittgenstein, the promoter of analytic philosophy. Some religiously minded folk want TV to be as soul-searching and soul-searing as Jesus of Nazareth. One may not, however, fairly expect a car to do what an airplane does. So, too, one may not fairly expect electronic media to do what only geniuses can do. Perhaps a closer look at TV's intrinsic limits can bring more realism into one's expectations about TV.

First, how many people semi-consciously expect to get as much relaxation from TV as from outdoor exercise--and with far less effort? Extended viewing of TV unavoidably reduces Americans' vitalizing encounters with nature since obviously TV sets are not designed to foster healing interactions with nature. Millions of Americans already work eight-hour days in wholly artificial atmospheres--that is, in offices, factories, and stores hermetically sealed off from nature and flooded with fluorescent light. What happens when, already deprived of so much contact with outdoor nature, they add to this unnatural constriction by daily spending four more hours in a closed semi-darkened room watching TV? Would not Americans benefit from nature's medicinal powers far more by walking and resting in a woods or meadow?

Again, if a person expects American TV to be ad-free, like some English TV, the result is frustration. For unlike the British Broadcasting Company, most American TV is controlled commercially and so cannot avoid making sales pitches one way or another. In order to survive financially, most TV broadcasters in the U.S. must lure their viewers consumeristically to a higher material life-style.

As wiser heads realize, America's genuinely happier people are usually far other than TV's "rich and famous" or its "bold and beautiful." Yet the glitter and glamor of TV advertising keeps luring its viewers to join the upper, economic class. This climb-higher lure is found in the commercials on TV, in newspaper and magazine ads, and also implied in most films. This lure often tends to make its viewers forget about the have-nots of the world. Sometimes such ads even nudge the thoughtless to look down unconsciously upon the many elderly folk and the many economically less advantaged of the world. In this way America's commercialized media convey to most women, Afro-Americans, and Hispanics the prejudicial message, "You're really not doing as well as the rich and famous." Thus an implicit sexism and racism is being broadcast through TV's enticement towards a materially higher life-style. Granted, this lure is currently an economic necessity in our free enterprise media system. To expect otherwise at present is to refuse to accept the media as they have to be here and now. For as long as America is a capitalistic system, the competitive forces between the advertising business corporations and the media need to be maintained.

Because the life of American commercial TV hangs upon advertising accounts, TV companies here are largely enslaved to their business patrons. Yet looking abroad to England, Germany, and the other largely socialist countries, media researchers see ways of at least reducing the ad clutter on TV here. English folk enjoy ad-free showings of programs like the popular Australian-produced "Neighbors." For when buying their TV set, they must purchase a license and renew it periodically. The revenues from these licenses largely support BBC's ad-free programming. In Germany viewers get many hours of ad-free TV by having all their advertisements clustered into half-hour broadcasts of nothing but ads. As a starter, at least, American networks could, like their cable competitors, charge monthly rental fees and free their programs from much of their current surfeit of advertising. But expecting this to happen tomorrow is unrealistic.

Again, some religiously inclined people expect TV to portray spiritual realities directly. One might think this about as realistic as expecting folk in Nigeria to live in igloos, but that would not quite be so. Movies like Stephen Spielberg's *Close Encounters of the Third Kind* or a TV serial like *Touched by an Angel* hint powerfully at spiritual realities. Yet scientific psychologists who dream that eventually TV will someday be able to show how minds work or how wills make free decisions are indeed dreaming. So, too, are any so-called theologians who dream of a day when TV will picture how saving grace touches hearts or even how the Holy Spirit works in the world. Such dreams, however, are fantasies that fail to take into account the intrinsic limits of TV.

Then, too, despite its efforts to allow its audience to be cool and laid back, TV is an intensely stimulating medium. Limited to being psychically manipulative, TV keeps on teasing viewers' attention forward. It keeps tugging at their interest, especially by flashing scenes of conflict or violence or by increasing the sound when airing commercials. It keeps seizing their attention by changing scenes, often quite rapidly, or by flaunting something new, unusual, or conflictive. If TV companies want to stay in business by maintaining their advertising accounts, they dare not lull their viewers to sleep. Instead, they have to focus on the fast-moving and startling, on flashy flip-overs, and attention-grabbing fights.

As a result, one may not fairly expect TV to portray the inner life of things that develop slowly. True, TV can portray the external growth of blossoming flowers by its trick camera techniques; but it cannot convey the real inner life of a plant. Much less can TV depict how a new idea or attitude slowly germinates, then surfaces, and later blossoms within the human psyche. Yet some people may let this shortcoming of TV lead them to believe that the only realities are those that people see and touch.

Furthermore, most mass media productions are geared to the educational level of sixth graders. This limits TV producers to make only gross contrasts--the kind that sixth graders grasp easily. As a result, TV producers can portray personality differences only crudely. They must make these differences seem as obvious as the white and black hats that distinguish the good and bad guys in old Western movies. Consequently, TV cannot do reflective questioning and critical thinking for its viewers. Yet alert creative minds will often press the muting button during TV's intermittent five to seven minute stretches of advertising. During these muted periods, such viewers can joke with or tease one another or begin their own critical comparisons of what they have just seen with other performances of the past.

Even when TV adhered to the practice of equal time, it was naive to expect TV to be politically neutral. It is even less so nowadays. TV has changed the whole way of doing politics in America. Simply recall from history how differently American politicians campaigned in 1900, in the early 1960's during the Kennedy-Nixon debates, and in the Clinton-Dole national election that spent well over hundreds of millions of dollars on political ads on TV.

Janet Hook and Sara Fritz, staffers of the *Los Angeles Times*, wrote in late 1996 that the Republican National Committee reported disbursing about $ 226 million in the 1995-96 cycle, while the Democratic National Committee reported disbursing $ 130 million in the same period. This almost doubled the $ 185 million both parties spent in the 1991-92 cycle. This does not include the additional $ 187 million spent by both parties together in "soft money"--contributions that ostensibly do not benefit any particular candidate.

James Bennet reported in the *New York Times* for November 13, 1996, that during the pre-election period from April 1 to October 31, 1996, a 30-second spot during the 6 p.m. news cost about $ 5,000 in Los Angeles, about $ 600 in Albuquerque. The two campaigns and the two major parties broadcast roughly 1,397 hours of commercial ad time (about 58 days worth) in the 75 major media markets nationwide. Of the 56,949,561 TV commercials broadcast during this period, 752,891 were political ads (about 1.3%). Almost one in five of these political ads were for a presidential candidate, the remainder being ads for city council races upwards to the Senate. Thus, TV has become the tool of politicians with big purses.

Moreover, TV has distracted both politicians and most Americans from arguing about the deeper thornier issues of domestic and foreign policy confronting the U.S. And if one watches the way many politicians eagerly seek to get into TV camera range, one realizes another change TV has brought about. Many politicians smile attractively before TV cameras and want to appear judicious before their home constituents. They feel that in this way they secure their position better than by researching issues and serving as statesman.

The U.S. President (or one of his administrative officials) uses televised press conferences to address all Americans. Today TV offers him a more persuasive and subtle means of political control. Heavy-handed executive orders or threats are no longer needed to lead or control Americans. Mass arrests, like those of Japanese-Americans in World War II or of Berkeley protesters during the Vietnam war are out of date. Today, thanks to TV's effect on the masses and despite commentators' prompt reaction remarks, the president can exert a widely unrecognized control within the minds of independent citizens. Thanks to TV's techniques and power, he can influence what topics Americans think about and largely how they personally will associate with the topics he has chosen for his address.[10] Such is one example of TV's power in politics.

Overall, then, Americans need to correct many expectations they may have about TV. It is neither genuinely relaxing, nor free from ads, sexism, or racism. It cannot convey important but delicate differences in character, morals, and religion, nor is it politically neutral.

The Reasons for a Mainly Positive Attitude Towards TV

After this survey of some unrealistic but widespread expectations about TV, we turn to the evidence calling for a mainly positive attitude towards this predominant form of the media. We are not calling for a blind enthusiastic beating of the big drum for TV. Of course, Americans need to mount a discerning guard against TV's many negative features which make far more than cartoonists refer to its tool as the "one-eyed monster" and to its common fare as "garbage." More Americans, however, need to recognize TV's potentials and actualize them creatively. In people's response to TV, perhaps the greater danger lies in focusing mainly on its negative aspects and thus in overlooking both its potentials and the call to use these creatively.

Factually, how many people are convinced of the great good TV can do? For instance, how many recognize what the American Psychological Association's study found about children between ages three to five?[11] When such children frequently watch *Sesame Street* or *Electric Co.*, they become more alert and discriminate more keenly than those watching other TV programs. Similarly, when such children viewed *Mr. Rogers*, they learned to be more empathetic and pro-social than their non-viewing peers.[12]

How many people appreciate the good that TV artists have already achieved? These people--TV producers, writers, and actors--have brought adults noteworthy historical documentaries, such as the outstanding series on the Civil War. and on World War II. The TV series *Australia* and the National Geographic Society's many scientific studies of animals in the wild have enriched many people. In the arts, exceptional performances by symphony orchestras and by troupes doing ballet and drama have elevated the spirits of millions. In addition, sports enthusiasts are regaled with events as spectacular as Wimbleton tennis and Superbowl football.

Thus TV has shown its potential to enhance people's lives and to tell stories in an authentic, fascinating, and educative way. It can also promote debate and rational discussion. One easily recalls the debates between various presidential candidates. Then there are those discussions furthered by various panels of experts--for instance, Bill Moyers series on *Genesis* in which Christian, Jewish, and Moslem experts interacted. People remember the drama of several hearings in Congress or the emphasis against wife beating created by the murder trial of O. J. Simpson. Certainly TV has won laurels in bringing people abreast of world news--often informing them only seconds after the news break occurs. In the sports world the instant replay offers viewers more entry into the art and the key patterns of a play, as well as help to the referees in their decisions.[13] Few have any difficulty remembering the political, religious, historical, and cultural extravaganzas that TV has brought into their living rooms. One need only recall such spectacles as the dancers singing atop the broken Berlin Wall, or John Paul II at Cherry Creek Park outside Denver, or the Tall Ships in New York Harbor during America's bicentennial, or the glamor of a night of Oscar or Emmy awards.

Beyond these achievements, we need to look forward to the blossoming forms of electronic media. Current TV sets will become obsolete in the face of high definition TV sets, especially when these are yoked to a personal computer. If within the limits of town-hall propriety, the producers of TV programs will permit genuinely interactive audience participation, they will transform TV into an effective instrument for creating a wider practical democracy in America's increasingly pluralistic society. Instead of reading books drawn from local library stacks, students and others will download from famous remote libraries books and articles encased in electronic memories and stored on the World Wide

Web. When state-of-the-art telescopes beyond Hubble's power become mounted on internationally governed space stations and linked electronically to the TV sets, data-recorders, and analyzers of scientists still bound to earth, astronomy promises to make stunning leaps. Diverse uses of TV in medical research will surely multiply for already TV is promoting genetic engineering and laser-precise medical techniques. These avenues, however, are only a few paths for estimating how the electronic media will soon enrich people's lives.

After examining the shortcomings of the media--TV's glut of trivia, its fostering of illiteracy, its catering to the masses--Charles Van Doren calls everyone to draw up a balance sheet of pros and cons on the media. Eventually he offers his own, showing his choice of a primarily positive attitude towards TV and the media in general. He writes:

> *Because of the media* we understand democracy better than almost anyone understood it a century ago. *Because of the media* we have a deeper distrust of war. Not deep enough, as yet, but the idea is very new for most people. Belief in the natural inferiority of certain kinds of other people--take your choice--does not so easily survive when the media continually remind us of our similarities to them.[14]

A balanced attitude toward TV, then, calls for a primary emphasis on its achievements and potentials with a sense of our responsibilities to use TV creatively. Secondarily, we also need great caution to get beyond the deleterious effects of some current TV programming. For TV directors and artists, the challenge is to fulfill TV's potential as entertainer and educator of the masses in the arts and sciences, in political life and technological breakthroughs, and even in simple but deep-level philosophy and religion.

TV's negative effects have been trumpeted and bemoaned by many--including such authors as Manders, Chomsky, Parenti, and Postman. Yet to overemphasize the obvious and hidden deleterious effects of TV tends to keep people from using TV's positive potentials creatively. The challenge of this book is to exercise constructive criticism and adopt innovative habits when using the media. In this way we will use TV and the other mass media to their full potential. We know how quickly the forward-facing media capitalize on new technological inventions.

In brief, can we become as inventive and flexible in dealing with the media as are those who seize upon the opportunities unleashed by these new electronic devices?

After this pair of introductory chapters, we can turn to the general kinds of media. However, most of them--especially TV, radio, newspapers and magazines--get their vital energy and most of their financial support from the all-pervasive influence of the advertising industry. So our first need is to get behind the electronic bits basic to advertising.

QUESTIONS FOR DISCUSSION

1. How have you experienced TV changing your attitudes or expectations or beliefs?

2. During the past decade, has the amount of your oral and written communication remained about the same, or has one increased noticeably? In either case, why so?

3. What steps led you to communicate orally in a more effective way?

4. When Neil Postman says, "Americans watch television but we have not yet reached the point where we watch ourselves watch it," what is he alerting us to?

5. Discuss the built-in limits of TV as you have experienced them.

6. Recall some examples of the lack of political neutrality on TV. How has release from the equal-time requirement affected TV's political neutrality?

7. In what positive ways has TV influenced your life?

8. Describe what you would consider a balanced attitude toward TV.

Chapter 3

TRUTH IN ADVERTISING

In an ideal world, advertisements via TV, print, or radio would present products in a truthful and balanced way, neither exaggerating the positive qualities nor hiding basic weaknesses. In the real world, such truth is difficult to find. To ensure that their products appear most attractive to the consumer, manufacturers pay large sums to advertising companies. So advertiser must find ways to meet manufactures' expectations for more sales and profit. They do so by exaggerating claims at times, by targeting the more likely purchasers, by using gimmicks, rebates and puffery, and, in general, by so presenting the various products that potential buyers draw erroneous conclusions.

General Effects of Advertising

Advertisers serve the general good by informing the public of new products and by providing instructions for the best use of new or improved merchandise. They help people choose at the best available price the product that will meet their needs. Through the efforts of manufacturers and advertisers the material life-style of most Americans has continued to move upward as they consume more and more of the earth's resources.

In less developed countries, the marketing of U.S. products has had some positive effects. Indirectly, advertisers educate these people to better nutritional and health care practices and to a wider global awareness. Yet advertising some products to these nations has had serious side effects. For example, in Latin America and Africa, advertisements for baby formulas as a substitute for mother's milk led to serious illness and even to many infant deaths. For without cautions to the contrary, many mothers in these countries mixed the formula with contaminated water, the only kind of water on hand.

Advertising has grown into such a giant business that in 1994 it had already reached sales of $150 billions annually.[1] It employs thousands of market analysts, artists, sales personnel, and support staff. In this way, it generates income that can be invested or circulated directly back into the economy through purchases.

The information in an ad may be true but be misunderstood by the consumer. For example, how many diet conscious people understand what "fat free" on a package of food really means? Some may believe they can eat such packaged food in unlimited quantities without gaining weight or increasing their count of bad cholesterol. The consumer can learn a great deal about a product and make intelligent choices by studying the dietary information currently required by law on the product's label, by checking the claims of advertisers and by understanding the vocabulary of commercials.

Many Americans may pride themselves on being astute and perceptive in reading ads and viewing commercials on TV. They may believe that they always find the truth in advertising. As a matter of fact, however, evidence shows that advertisers use more subtle ways of appealing to the public than most Americans even imagine. Things are not always the way they appear on TV. For example, a commercial intended to show the strength of Volvo's car roof showed a pickup truck with giant tires driving on the top of a row of cars. All the car roofs collapsed except that of the Volvo. What the ad did not show was the special braces that reinforced the Volvo's roof for the ad. The automaker had knowingly advertised advantages that the car did not possess.[2]

Commercials for miracle diet pills can mislead a gullible person desperate to lose weight. One ad that was challenged claimed that a person could "lose 6 to 30 inches on the first visit" to a diet clinic. The Federal Trade Commission (FTC) has brought numerous suits against false advertising for weight reduction products but there is never a shortage of such ads.[3]

These ads are not the only false commercials investigated by the FTC. Recently the FTC announced settlement with two New York City law firms which ran deceptive advertisements for credit repair services by posting these ads on 3,000 Internet news groups. The lawyers claimed they could restore a good credit rating to people who had a poor credit standing. They referred to themselves as a Consumer Credit

Advocacy (CCA) group supported by experienced attorneys. According to the FTC, the law allows a firm to report to companies accurate up-to-date credit information. Derogatory information, if true, cannot be removed from a credit report.[4] Fraudulent ads, such as the CCA circulated, diminish the trust of consumers in all commercials and undermine the value ads can have.

As we shall see later in Chapters 7 and 8, the media find themselves pressured in many ways and, in turn, influence many attitudes and the life style of countless Americans. As Harry James Cargar who has analyzed the impact of TV violence on the audience, remarks:

> The values that advertising depends on and encourages have to do with acquiring, with owning, with obliterating frustrations, with pleasure now. These indoctrinations, when thwarted, can lead to a sense of failure, to disappointment, to revenge and physical violence.[5]

Image Makers

Already in the 1970's, Madison Avenue advertisers realized that a whole segment of Americans opposed the consumer culture. About fifty million disillusioned people between the ages of twenty-five and forty seemingly rejected the values of the materialistic culture and began searching for personal fulfillment rather than money and luxury. In order to get these people purchasing again, advertisers employed psychologists and sociologists to research the vulnerabilities of various segments of society and to find easy remedies for their anxieties and insecurities. This research resulted in changing the emphasis in ads from how well the brand performed to how the customer feels about the product. This type of research led to psychographic approaches in advertising, of which the most widely used today is Values and Life-Styles (VALS), designed by SRI International (formerly the Stanford Research Institute.)

VALS divides all citizens of the United States into five groups:

1) The Belongers value security, the status quo and a structured society. They feel uncomfortable with change. Commercials aimed at this group offer a world of idealized images featuring family activities.
2) The Emulators consist of young people searching for their identity and their place in the world. They respond to commercials showing models--macho men and glamorous women--using a particular

product and all saying in effect, "Use this product and you'll be like us."

3) The Emulators-Achievers might be called the "acquisitive consumers." On the way up the ladder of success, they fear they will not reach the top rung. They become compulsive consumers being trapped by commercials that promise them the look and feel of the wealthy and most successful.
4) The Societally Conscious Achievers offer advertisers their greatest challenge. For this group, personal fulfillment matters more than climbing the corporate ladder. These desire fitness and health and respond to commercials carrying an image of simplicity and integrity.
5) The Need-Directed, the poor and lower middle class, spend most of their money on necessities. Their lack of money makes this group practically invisible to advertisers.[6]

Having identified the four groups who spend more freely, social psychologists undertook extensive interviews to pinpoint brands of products used by the majority of people in each group. For example, which cereals do Belongers choose? Which cars do the Emulators drive? Which beer appeals to Emulator-Achievers? Having gained this information, advertisers can create an image of a product to appeal to a specified group.

When the sales of a product show a decline, the image-makers become very active . For example, busy dog owners liked Gainesburgers dog food because its burgers were individually wrapped. Gainesburgers sold well in the 1960's even though it cost more than canned food. In the recession in 1973, however, the drop in sales required a new image to pull dog food purchasers away from the cheaper brands. Research showed that Belongers love their dogs. So the new commercials showed a dog doing something special--bringing in the paper or fetching dad's slippers. Meanwhile the announcer told viewers that in return their dogs deserved loving care via special food--Gainesburgers--even if it cost more.[7]

Gimmicks and Rebates

A trip to the supermarket can demolish the budget of a person with the tendency to spend readily. So many gimmicks stimulate people to buy on impulse. One supermarket experimented with a soft-drink counter dispensing a small free Coke. A survey showed that 80% of the people accepted the Coke and then spent $2.44 more than the average customer formerly spent.

Seeing the number of mothers who bring small children to the supermarket, another store increased its sales by providing small carts for the children to push--just like their mothers. Rather than risk a temper tantrum at the check-out counter, the mothers paid for all the items their children had collected. Furthermore, most supermarkets keep the shelves full since studies revealed that customers buy 22% more if the shelves are kept well stocked.[8]

Some manufacturers use rebates as a subtle come-on to sell products to any of the VALS groups. Instead of decreasing the price by a certain amount, manufacturers charge a higher price and then promise a rebate. The manufacturers want to maintain their profit so they find ways to make collecting the rebates possible but difficult. For example, Valvoline promised a $2.40 rebate with the purchase of 12 quarts of motor oil. The usual oil change takes four or five quarts of oil, but to get the rebate the purchaser had to send in the sales receipt with the emblems from seven cardboard containers within thirty days. Since the cardboard containers often leaked after being stripped of the emblems, collecting the rebate became very difficult, if not impossible.

In a survey on rebates, a woman reported that she liked a cup of tea in the afternoon and had bought Salada Tea after she noticed she could get a $1.00 rebate on it. She discovered that the details for obtaining the rebate were on the back of the sticker. She had to open the box to read the sticker which revealed that she needed two boxes for the rebate. She also fared no better with the carton of Mott's apple-grape juice which promised a free carton with proof of purchase. In this case, proof of purchase consisted of price codes from five cartons.[9]

One gimmick has mushroomed in recent years; the grocery coupon for easing purchase of brand-name foods in a supermarket's other aisles than those that offer of green produce. For instance, Natural Cereal, priced at $4.75 per box, can be purchased at $3.75 with a coupon. For a particular product like Natural Cereal thousands of coupons are published. At least they serve as widespread ads even if the vast majority are never used but discarded with the daily newspaper. Advertisers have researched just what percentage of purchasers will buy Natural Cereal, without using a coupon--often busy mothers, or husbands returning from work too tired to bother with coupons. Whenever a customer buys Natural Cereal this way, he or she contributes $1.00 to the manufacturer's profit. Advertisers also have researched what percentage of purchasers will take time to clip coupons, store them for the shopping day, and then take care while shopping to buy the brand for which they have coupons. Coupons seem to offer savings but certainly don't decrease the manufacturers' profits.

Puffery

The success appeal of advertising puts such a strong stress on upward mobility that few can resist the lure. The consequences of succumbing to the attraction of moving up in the world can be serious. The Simmons, a young couple, discovered this when they bought their first house. They could not resist the ads featuring new homes in a developing area on Long Island.

The purchase of a home represented a big investment for the Simmons family and Harry tried to be responsible. The Realtor told Harry he had just the place for him in South Riverbend. The Realtor took Harry and Mary by the scenic route to see the model house. Mary was fascinated with the mirrors on the walls in several rooms, and the bedrooms looked especially large even considering the fact that some furniture seemed to be missing. The bath off the master bedroom had a whirlpool and the swimming pool in the back yard attracted both Harry and Mary. They never imagined they could own such a beautiful home in such a high-class neighborhood. They were so excited, especially when the Realtor told them the house would be ready for them in four weeks.

Their first disappointment came at the closing when they discovered that the swimming pool was not included in the quoted cost. By now they loved the house so much that Harry said, "Well, I'll figure out some way to make the higher payments."

They had sold their house and packed to move on the promised date only to find that the house would not be ready for another six weeks. The Realtor was vague about the delay and they began to suspect he had only promised the early date in order to make the sale.

Finally, moving-day came and with it another unforeseen disappointment. While the rooms with the mirrors seemed large, they had a hard time arranging their furniture and when everything was in place the room seemed very crowded. They never discovered that the furniture in the rooms for display had been scaled down and so gave the impression of more floor space than was actually available.

Harry's frustration knew no bounds when he tried to get on the I-495 expressway to go to work. He had been led to believe that South

Riverbend shared the advertised conveniences of Riverbend. Now he discovered that this was only partially true. He could find no easy and direct way to the westbound ramp of I-495. As a result, his commuting time to work almost doubled.

Harry and Mary were victims of what is known in the building industry as "puffery." Puffery violates ethics when it intentionally leads buyers to false conclusions, but victims find it almost impossible to prove in court that the seller intended the false conclusions.

A classic scam that fooled many people in the past involved the sale of lots in the swamps of Florida. One Realtor was more clever but no more honest in his project. In his brochure, he emphasized all the amenities that would be available in the sub-division under construction. He sold 300 homes but the owners found none of the promised amenities. To get zoning for the sub-division, the builder had promised City Council he would construct a less-expensive, more affordable section as soon as the amenities were completed in the 300 homes already built. By delaying the amenities as long as possible, he put off fulfilling his obligation for constructing the more affordable housing.[10]

Being aware of the hype and exaggeration of ads does not always protect readers and viewers from sometimes getting caught in the habit of the quick, unthinking response. After an unfortunate experience with the puffery of the real estate business, a young man thought he was looking at all ads with a critical eye. Yet when he read on the cover of *Life*, Nov. 1990, the claim, "After 16 years, an offer no movie-goer can refuse . . . GODFATHER III," his immediate reaction was, "Like hell, I can't!" It did not occur to him that in trying to protect himself, he was forming the habit of quickly tuning out all incoming impressions without critical evaluation.

Product Presentation

Since the Federal Communications Commission lifted restrictions on the time limit of commercials in 1984, a new form of commercial has emerged aimed at the late night television viewers. These commercials, called "infomercials," seem to the unwary viewer information about a type of product. The format resembles that of a talk show, a half hour or even an hour long, complete with host and a number of guests. At this late hour many viewers are not sufficiently alert to realize that

1) the health cures, or sure-fire diet programs, or the money schemes are too good to be true;
2) all the commercials within the program promote the product of only one manufacturer;
3) all the guests praise only one product, never raising critical questions or pointing out any negative aspects;
4) nothing identifies this program as a paid commercial.[11]

Very similar to the infomercials on TV are the advertorials in magazines. Advertisers use advertorials as a means to draw attention to their product within the supportive editorial of a magazine. The advertorial is designed in the format of an editorial and may easily be mistaken for regular editorial matter rather than the promotion of a product. For example, an editorial on safe driving may contain ads featuring the safety features of a particular car adjacent to the editorial. The advertorials blend so well with the rest of the magazine that the intent is not clear unless the pages are marked "Special Advertising Section." More and more advertisers demand control over the editorial matter connected with their ads. Some magazine publishers fear that control of editorials will threaten the magazine's integrity with their audience.[12]

According to Jean Kilbourne the primary purpose of the mass media is to make audiences available to the advertisers. Alcohol advertisers have created some common myths in order to attract the audience:

1) Drinking is a risk free activity. There is never any sign of trouble or addiction revealed in the ads.
2) Ads make the celebration of drinking itself--not a holiday, festivity or family event --a reason for drinking.
3). Slogans like "your own special island," or "your mountain hideaway," encourage isolation. The ads do not warn of in-reasing alienation.
4). Alcohol is advertised as the magic potion that can transform you. It is linked to happiness, wealth, prestige, sophistication, success, maturity, athletic ability, sexual satisfaction--everything that alcohol's misuse will destroy.
5) Sports and alcohol go together. The truth is that alcohol decreases athletic performance.

6) The naive claim that if these products were dangerous the media would warn the public. They fail to note the statistics: Each year illegal drugs are linked to 20,000 deaths, alcohol to 100,000 deaths, and cigarettes to 450,000 deaths or over 1,000 a day.
7) Companies that make alcoholic beverages promote the slogan, "Don't drive when drunk." They imply it is all right to get drunk if you just don't drive.[13]

There seems to be no end to clever ways of advertising a product. The rapid growth of CompuServe and America Online is the result of shrewd deals made with newspapers and magazines to get exclusive access to their copyrighted material. In September, 1993, *Time* began putting the full text of its weekly edition on America Online. As a part of the deal, *Time* ran six free, full-page ads for the computer service. In this way the ads reached millions of readers and brought 10,000 new subscribers to America Online.[14]

Marketing by Mail

Americans are bombarded by ever increasing ads in newspapers, magazines, and direct mail to their homes. Thus there seems no limit to the psychological drag toward "moreness" in our lives. This drag often influences unconsciously those who feel they resist ad appeal. Popular TV series flaunt lavish apartments and homes where families live in luxury and strive for nothing except more wealth, more power, and greater personal happiness. Anthropologists Mary Douglas and Karen Wollaeger call such serials "dramas of disruption."[15] As portrayed in the serials, the disruptions make fraud, infidelity, and divorce attractive and thus defy and corrupt traditional norms. Week after week they repeat the message: "The wealthy are immune from the consequences of their deeds."

In some cases, the fast pace of American life acts as a saving barrier against this consumerism since people do not have time even to glance at the vast amount of print advertising. The unsolicited catalogues, brochures, and flyers only add to the pollution problem. Enough people, however, seem to read printed ads, at least superficially; otherwise, companies would not find it profitable to continue this form of advertising. Many people, unfortunately, do not have the time to read the small print and sometimes find they have unwittingly placed unwanted orders.

People who install a swimming pool or even move into a house with an existing pool are surprised to find in the mailbox brochures like the one from Sundancer in Boca Raton, Florida. The brochure contained pictures of 18 swimsuits with the message:

> We are conducting a test to choose which of our bathing suits to introduce in the United States, Europe, and the Orient. Before we take this giant step, we need your help. Please look over the enclosed brochure and just tell us which of the suits you would choose. Give us your first and second choices.[16]

A form was provided for marking the choices. After an unsuspecting, cooperative person marked her choices and returned the "ballot," she was surprised when the swimsuits arrived several weeks later. When she tried to return them, she discovered that the place indicated to mark the choices was really an order blank.[17]

Schemes such as this have given direct marketing by mail a bad name. Some companies have suggested self-regulation by an ethics committee to investigate these problems and negotiate solutions. Terrance H. Witkowski, writing in *Marketing News*, contends that:

> The ethical issues that will eventually hurt the industry include: promotional excess that bothers consumers; unfair and aggressive advertising that takes advantage of certain consumer groups; misleading and deceptive copy and envelopes.

He strongly suggests a ban on automatic dialed recorded messages and a reduction in the over-circulation of name lists.[18]

Owners of small businesses often become victims of advertising schemes because they think they know the tricks of the trade. When Sylvia, the owner of a small dress shop, read the offering by Wells Fargo Bank to insure credit cards against loss or theft, it seemed like a good idea for someone who traveled as much as she did. The ad claimed that for $12.00 per year Hotline Credit Card Protection would reimburse the holder up to $5,000 per incident. Sylvia did not realize that California law requires sellers of credit-card insurance to notify prospective buyers that under federal regulations a owner's liability for charges on a stolen card is limited to $50.00. When challenged, Wells Fargo Bank pointed out that the required notification was on the inside surface of the envelope--"strictly a layout decision" according to one bank representative.[19]

Exchange of lists of addresses can get out of hand. The editor of *Consumer Report* ordered a $60 electronic gadget from Sharper Image in May, 1993. Within six months he got three more catalogues from Sharper Image besides nineteen other gadgetry catalogues. The mails are flooded by speciality catalogues sent to small groups who show interest in a sport or hobby by subscribing to a particular magazine. Many catalogues gush like the salesperson who tells you everything you try on is marvelous. For example, according to the catalogue, Thompson's shirts "may be one of most visually interesting shirts you'll ever wear." Lands' End advertised shorts by featuring short stories about shorts. Another gimmick is the use of "only" and "just" before prices when actually other companies charge less. Catalogue copywriters compete to come up with obscure names for colors, such as "mountain green" or "butter yellow," thus guaranteeing that you'll never create a matching outfit unless you order another item from the same company. The newest version of mail order marketing allows the consumer to shop electronically using TV, CD-Roms and on-line computer services rather than a traditional catalogue. Estimates indicate that by the end of the 1990's this billion dollar industry will grow to $20 billion.

Quick Riches

Night after night, a young woman named Carol, watched the ads for the lottery. Every Saturday she was glued to her television set with tickets in hand, checking whether she had won. Then one day, it happened--Carol won BIG! Carol became ecstatic and her friends rejoiced with her. Carol figured that over the years she had invested more than $12,000 in the lottery and now she felt repaid. Even before receiving the first check, she was deluged with circulars, ads, and phone calls promoting expensive furniture, jewelry, cars, and boats. She changed her phone number to an unlisted number, but her doorbell rang constantly and she feared to go out. She kept her daughter Sally at home with her, afraid to let her go to school. "She might be kidnaped," Carol said to her husband. The family decided to move to Oregon to escape all the publicity. Unfortunately, the Realtor selling the house they wanted to buy discovered their secret. The price of the house jumped overnight from $100,000 to $169,000. Rumors spread through the little town, and then, far from home, and away from their friends, the harassment began again. Recently, Carol wrote to an old friend, "I wish I had my old life back."

Carol is just one of many lottery winners who discover that their lives are changed in ways they never imagined after they win the big prize. Their friends drift away because they do not have the money to associate with someone in a fashionable neighborhood. The winners are not accustomed to their sudden wealth and new life-style. They feel uncertain and insecure. Relatives often expect them to share their good fortune and make comparisons that cause rifts in the family. Besides the problems that winners experience, the lottery creates more subtle and often unrecognized or unacknowledged difficulties for those who play every week for years and never win. Lottery ads are aimed at the people who can least afford to throw away their money. These ads create a craving for instant, unearned wealth. The longer this craving goes unsatisfied, the more intense it becomes. An unhealthy greed replaces the healthy ambition to work hard and enjoy the fruits of one's labors. Frustrated greed can lead to creative but illegal ways of obtaining what the lottery fails to provide for the thousands of disappointed players. If a steady loser turns to illegal ways of getting money to satisfy frustrated greed, he or she resembles the person who pushes drugs to win quick easy riches.

Sex and Violence in Ads

Since TV advertising has its own game plan, woe to the unwary who tries getting an ad on TV for the first time. One soon discovers that ad executives have one criterion: promote only those ads that will entice most customers. This criterion translates for these ad executives into "Use sex or violence, or both."

On the other hand, network executives show somewhat more caution for fear of evoking public criticism or offending their more important customers. They protect their own vested interests. A constant battle wages among the networks for the largest audience as expressed in the highest ratings, thus attracting more advertisers at higher and higher rates. This desire for ever increasing profits leads to under-representing women and minorities and downplaying thoughtful or specialized programs. For sex and violence have proven to be the attention-getters that draw large audiences. As the media expert Benjamin Bagdikian puts it, "The intensity of these two elements is increasing not only in the U.S., but also around the world with the spread of uninhibited commercial TV which needs attention-getting devices in order to command the highest payments for their commercials."[20]

Many people, especially women, agree with Rachael Lauer when she says:

> Exploiting our sexual responsiveness by associating it with commercial products, I maintain, is an attack upon our humanity because there is an inherent cynical trivialization of sex when associated with cigarettes, wine, soap, etc. A fully human person does not respond automatically to stimuli but chooses. It is a waste of energy to be forced to resist these daily bombardments. The best criterion for evaluating sex in advertising is not by the degree of sexiness but the degree to which it allows us to use those powers that signify our highest humanity, the powers to reflect upon and choose what goes into our consciousness.[21]

When faced with protests about the amount of sex and violence on the screen, TV executives lamely excuse themselves, "We give the American viewers what they are looking for." Yet recent statistics seem to contradict this claim. A survey of 4,000 people by DDH Needhame Worldwide Inc., New York, found that 72% of males and 84% of females believe that television advertising puts too much emphasis on sex. Furthermore, Marty Horn, vice president of this firm's Chicago office, says the percentages of this survey have been stable for the last ten years.[22] Moreover, as recent studies show, publicity about sexually transmitted diseases--particularly about AIDS--has led advertisers to down play sex in most advertising. Large advertisers of beer and cars are reducing the sexual tone in their ads. To take advantage of our aging baby-boom generation, conservative ads, built on images of family, home, and security, are becoming commonplace.

Nevertheless, specialized advertising still uses sex extensively in commercials for perfumes and fashion items. Specialized magazines that target a trendy, upscale market are noted for their sexual tone. Some writers, such as Andrew Sullivan, use strong language to express their evaluation of fashion advertising. Sullivan maintains:

> Fashion advertising has been infected with a brazen new elitism, sexism and objectification of flesh. Abuse of women, homosexual erotica and the milder forms of sadomasochism are now commonplace marketing techniques directed to the average American household. Products as everyday as salami, chemicals, and pianos are being sold with sexual explicitness once reserved for paying customers of soft-core pornography. Perhaps the best adjective to describe the ethic of the whole power-sex genre of advertising is "fascist." For example, Calvin Klein's "Obsession for Men" shows four men and two women, naked

and draped over a Nuremberg-style obelisk, looking off into the distance. The faces in this ad are blank, giving no evidence of caring for anything or anyone, not communicating, but just staring off into space. Perhaps the most striking feature is the dehumanized uniformity of bodies they display.[23]

Subliminal and Surreal Advertising

In the early 1970's, Wilson Brian Key created a stir by his book *Subliminal Seduction.* He claimed that messages were transmitted by fleeting images which, although not seen consciously, created a subconscious impression. In 1972, the FCC investigated subliminal material used at drive-in theaters. Fleeting images of popcorn flashed across the screen to encourage patrons to buy the snack. The patrons were not consciously aware of the source of their desire for popcorn.

It was later claimed subliminal material was not effective since the stimulus was too weak; as a result, the use of such material seemed to decline. Key maintains, however, that covert messages are being used with greater intensity than ever before.

Key's conviction finds support in the use of subliminal messages on computer screen savers. Screen Team in Dana Point, CA, offers advertisers the opportunity to incorporate subliminal messages on screen savers as well as on word processor and spreadsheet programs. This company is also considering marketing some products that would help consumers quit smoking, think positively, or lose weight.[24] David Pescovitz, writing in *Wired,* reports on the discussion of advertising experts about the future of advertising. Their prediction for the future includes subliminal advertising and also advertising in cyberspace. They believe that a ban on using unsolicited E-mail to advertise is unlikely.[25]

Shelly Palmer, a copywriter, singer, and sound guru, experiments with another form of subliminal material. By altering the sound wave of voice and music track in ads for Seagram's mixers, he subtly mimics the sound of ice cracking. Since FCC rules prohibit ads depicting consumption of alcoholic products, the sound of ice cracking evokes the emotional response that vision cannot provide. The FCC also bans the airing of sounds of gambling equipment in ads for bars, nightclubs, and casinos. Palmer has produced an ad with subliminal sounds of clanging, clanking, and bell-ringing to accompany commercials for Bally's Casino.[26]

Much research has been done on the effectiveness of non-verbal communication in advertising, primarily in the area of music. There are basically two overt ways that music is used--as the melody for a ditty and as background accompaniment for the commercial. A catchy tune has proven effective as people find the tune re-echoing in their minds as they go about their daily activities, walk down the street, or ride to work. In a sense, they take the commercial with them.

Studies of the second way to use music in ads--background music during a commercial--indicate that the effects are ambivalent. The music can be distracting, causing the viewer to give more attention to the music than to the commercial. The intent of the advertisers is to create in the consumer a positive emotional response to the product. There are so many factors involved, such as major or minor mode, tempo, rhythm, loudness or softness of the music. Preferences of the viewers to certain kinds of music make the task of matching music and commercial to create the necessary appeal very difficult.

The use of surrealism in advertising attempts to overcome the objections and inadequacies of subliminal ads. Surrealism is an art form of the twentieth century through which artists try to represent and interpret dreams and similar experiences. According to Murray and Murray:

> The object was to free artists from the normal association of pictorial ideas and from all accepted means of expression, in order that they might create according to the irrational dictates of their subconscious mind and vision. They seek to fuse waking and dreaming mental images.[27]

Surreal advertisements are aimed at influencing the consumer on a subconscious level, much as subliminal ads do, but with a very important difference. Surreal advertisements have a much greater stimulus intensity and sufficient visibility to overcome the distrust associated with subliminal advertising.[28]

Surreal advertisements are another means to create the unreal world described by Joe Saltzman:

> The world of TV commercials is one in which airplanes never crash, cars never break down, all food is appetizing, no dishes or floors or toilets are ever dirty, no baby damp, no woman or man is unkempt or bad smelling and nothing is bad for your health or your life style, everything looks real, and feels real but isn't real at all.[29]

Next Generation

The power of advertising affects very young children especially. The Saturday morning cartoons are interspersed with commercials aimed at the young viewers, creating in them desires for expensive toys and video games. The fact that some of the toys are "educational" does not justify the effect of beginning to mold these budding personalities to be acquisitive. Even at an early age children are encouraged to want more toys by the abundance of new possibilities flooding the market, especially before Christmas.

Advertisers are well aware that there are 54 million children between the ages of 3 and 17 with money to spend. Besides their own money from gifts, allowances, or pay for small jobs, which totals $104 billion, they strongly influence about $160 billion of the money that their parents spend annually. Advertisers are, therefore, very motivated to attract a large share of these billions. It has been estimated that children view 30,000 ads each year, more and more of them in school. Because many schools are seriously underfunded, teachers and administrators are pressured to accept ads disguised as educational materials such as McDonald's "All-American Reading Challenge." The ad-sponsored daily classroom news program, Channel One, claims to reach 8 million students in grades 6-12 with its news and two minutes of commercials. Among the most harmful advertising for children are the so-called packets that claim to teach methods of scientific inquiry. The children are sometimes led to false conclusions for the benefit of the sponsoring company. For instance, the National Live Stock and Meat Board is understandably interested in increasing the consumption of meat. One of its packets for schools leads the children to infer that early American pioneers were short because they did not eat enough meat.[30]

Many of the video games that have become so popular on computers, in homes, in malls, and in theater foyers encourage a competitive spirit and the illusion of a quick solution if one just makes the right moves promptly. Video games reward the quick eye, manual dexterity, and impulsive reactions while leaving no place for patient reflection or critical thinking. How many parents even suspect that they may be submitting their children to cybernetic engineering? Do they want the brains and nervous systems of their children mainly to react automatically to mechanical-electrical communication systems?

The next generation may well become what William Donnelly calls the "Confetti Generation." For they may develop a "deep psychological depression bred of inner confusion and weightlessness." Whatever is perceived will be seen as equal. When religions, life-styles, and all other ideas are equally valid, then they also become equally indifferent and equally undifferentiated. Their value will depend merely on the choice of the individual. Like confetti, all events, ideas, and values will be the same size and the same weight.[31] If parents allow the personalities of their children to be formed in this way, where will we find genuine thinkers and decision-makers for the next generation?

Unfortunately, the problem of maintaining truth and balance in advertising is not the only aspect of the media that influences our lives, sometimes in a negative way. As we shall see in the next chapter, media executives, editors, and producers face almost insurmountable problems in the selection and balance of the news--whether local, national, or international.

QUESTIONS FOR DISCUSSION

1. In which of the five VALS groups do you belong? What kind of ads should you expect in that group?

2. When shopping at the supermarket, do you bring a list and stick to it? What are the benefits of planned shopping?

3. Describe your experience in collecting rebates or using savings coupons.

4. Have you or a friend ever been a victim of "puffery" or an advertising scam? Describe that experience.

5. When hit by unsolicited marketing phone calls, how do you react?

6. What makes the poor especially tempted by lotteries and casinos?

7. Describe the changes occurring in the tone of advertising in the past five or ten years.

8. What effects have you noticed in children who spend hours playing video games?

9. Do you believe in subliminal messages in movies, recordings, and ads? If so, what evidence do you have to support your belief?

Chapter 4

SELECTION AND BALANCE OF THE NEWS

The previous survey of searching for the truth in and about advertising is now followed by one of the media's more attractive functions: reporting the news. Editors, news writers, reporters and camera crews have a very challenging vocation: to serve the general public with news that is factual, interpreted fairly, and reported in a balanced way. Given the constraints under which these people labor, many of them turn in amazingly competent performances. Television newscasts today display the awesome power of news personnel to collect and present a vast variety of news items drawn from around the world. They do this in an informative and interesting way, within the limits of rigorous editorial decisions and deadlines, and usually within a span of no more than twenty-two minutes during a half-hour newscast.

Then, too, many TV news anchors come across as pleasant companions to whom living room doors would readily open. One thinks of Tom Brokaw or remembers avuncular Walter Cronkite. Many anchorpersons maintain a light-hearted banter that offsets the sometimes tragic news or cruel scenes they report. They thus inject a sense of life's ongoing flow and saving humor to achieve some kind of balance.

Furthermore, if one remembers the news personnel who have been killed trying to film or report stories, one realizes the courage that is often at the heart of a dedicated news person. The fourth estate certainly has its heroes to whom every media-user is indebted, even if we have not as yet established a national memorial to honor courageous reporters. In brief, a strong case can be made that some of the best work in the media is done by those who produce our newscasts.

Nonetheless, how the world looks and sounds on TV news programs depends upon which events have been artfully distilled, selected, and framed into this tightly packaged composition. Hence the need to penetrate to the actual events behind the scenes and sound bites so carefully packaged for these newscasts. As more and more Americans look to TV news for information, a common question that critics raise is, "Which people, places, and happenings are chosen to represent the reality of the day?"[1] How might TV news distort the shape of reality?

In Chapters 7, 8, and 9, we will learn how the media influence Americans' lives and how the increasing concentration of communication corporations verges on a monopoly.[2] Among the many ways in which this influence and near monopolization express themselves are the selection and sometimes biased treatment of the news. Accordingly, if the present chapter focuses on some ways of slanting the news, it can only provide a partial picture of the overall power of the media at work. Again, this chapter might also be misinterpreted as suggesting that selection and balance in the news are separate operations. Factually, however, these two overlap so much that it seems more honest to view both of them as arising from an intent that is overly economic and thus manipulative of how people think.

Consequently, this chapter aims first to appreciate how selective editors must be and how difficult it is to provide a balanced treatment of the news. Then it will focus on slanted selection of the news--that is, such selection as puts an unfair interpretation on the news. It will conclude with instances of excessively unbalanced treatment of the news.

Selection of the News

As Gregor Goethals sees it:

> [To] comprehend the authoritative power of news programs we need to be conscious of the creative, selective process involved in the re-presentation of events that flow across the TV screen. Images that seem to put us directly in touch with distant places and events only appear because control--both technical and editorial--has been exerted at several levels.[3]

In response, media owners and editors will, of course, issue their disclaimers. Some students of the media, however, speak out against the media bias that arises from the dominant operating intent of media owners and editors. As for-profit entrepreneurs, they *must* satisfy stockholders' yen for more return on their investments and *must* retain advertising accounts. A recent microeconomic study reveals how the market shapes commercial news production.[4] Not just the consumer market but the markets for investors, for ads, and for other economic factors press in upon editors and shape their decisions whether to deploy reporters, what kinds of resources to send with them, which stories to select, and which ones get classified as news. The vested interests of media owners and editors exert a control on the selection of news items and exercise a power felt throughout the media industry. Large advertisers and public relations firms concur in exerting parallel pressures because of their financial interests.[5]

Perhaps a lesson from history may begin justifying the claim just made that the intention of media owners and directors is excessively profit-centered and thus manipulative. At the start of the 19th. century, long-distance signaling enabled Nathan Rothschild to make the quick moves that led him to amass much of his fortune. Before 1810, Nathan had established in Europe an unparalleled news-gathering service through a system that flashed signals from one station through a series of intervening hilltop stations to headquarters. So when Napoleon's defeat at Waterloo occurred in 1815, Nathan's signaling system made him the first financier to be informed of the debacle. At once, he sold heavily and, by so doing, intentionally initiated a financial panic. This soon allowed him to pick up sharply devalued properties quite cheaply. By this shrewd quick use of his long-distance signaling system, he rapidly enhanced his fortune.[6]

Once Marconi had invented the telegraph, it too became a tool used with entrepreneurial intent. Today, wireless and the later instruments of telecommunication--radio, wire-services, television, satellite systems, faxes, and e-mail--function as adroit tools for generating capital. As a result, when media editors employ such profit-making tools, they are co-opted into selecting and treating the news in ways that usually fit in with the interests of big business, for reasons explained in chapter 9. In this way the majority of the human race, whose main interests are simply survival and basic human growth, are subordinated to the primacy of capital, the value priority of big business.

An instance of this entrepreneurial bias occurred in the coverage which the media gave to the Watergate scandal. The media focused their spotlights, not on the convicted business people involved, but on the political figures: John Mitchell, H. R. Haldeman, John Dean, and John Erlichman. Yet twenty-three leading businessmen were convicted of Watergate crimes: Claude Wild, William Keeler, Orin Atkins, and twenty others.[7] Why did the media not cast an equal spotlight on at least some of these businessmen? Why, too, did these involved businessmen find their sentences suspended? If fair-handed treatment is a media ideal, why did the media not pursue them as unflaggingly as it did the four political figures?

Another hint supporting our claim of biased selection comes from Landrum Bolling of Georgetown University's Institute for the Study of Diplomacy and editor of *Reporters under Fire*. Although his frank remark focuses on newspapers and news magazines, his charge need not be restricted to these media only:

> For good or ill, newspaper readers have been, for generations, the captive audiences for some pretty opinionated editors. And not always have their opinions been confined to the editorial columns. The weekly news magazines, moreover, have tended to make no separation between editorializing and straight reporting.[8]

Consider for a moment, how you would feel if you ordered a main course in a restaurant and in response the waiter brought you a glass of sparkling water. You would feel cheated, and rightly so. Many people, however, order the full story from TV news yet do not realize that sometimes they are being offered just sparkling water instead. Despite the vast technological infra-structure and the complex achievement of news reports, the media's news stories, especially those on TV, are highly selective.

Constraints Requiring Selection

When one turns to a news broadcast, he or she may want the facts and nothing but the facts. Yet media people labor under many constraints--constraints imposed by the nature of the media themselves and by their audience. A look at both these kinds of constraints may draw attention to omitted types of items and to innuendos that editors actually, but perhaps unintentionally, make.

For instance, what types of items <u>do not</u> get into a TV newscast? Events simply reported to newsrooms by a non-professional in the news world will not make TV news. So, if an attentive passerby, or some housewife or a neighborhood kid calls in a significant story as it breaks, that will not make the newscast by itself. With their felt responsibility to the general public, news editors rightly prefer items properly identified by reliable news reporters or sources.

Moreover, if camera crews cannot get to an event, it will not become an item on TV news. Or, if the event lacks action and drama, it will usually be excluded. Only an unusual and notable event, professionally reported, counts as news. So, the fact that a mother cooks another supper on an ordinary day or that a sales clerk courteously serves a shopper does not qualify as news.

Selectivity of the news may sometimes arise when editorial teams have been entrenched in the same offices that they have occupied for decades. Self-perpetuating editorial teams often lead to a stagnated viewpoint that automatically eliminates different perspectives. So, when high media positions are awarded undemocratically, a selective mind set is often imposed on the public.

Newscasts have other constraints that require even greater selectivity. Besides the topic selected for presentation, news workers have schedules to meet and editorial clearances to pass. They have to meet the technological demands of rapidly collecting, processing, editing, and presenting the news in an attractive way. They also have to target the news releases so that they fit the times when most of their audience can receive their broadcast--e.g., around dinner time or during home-bound commuting. All these constraints on media personnel require the exercise of a high selectivity toward news items that might be broadcast.

Then, too, editors must judge how interesting an event is for their intended audience. To the audience in a particular American city, one murder there will be far more interesting than the ongoing starvation of a million African children on the same day. Sometimes the TV news will even pass over today's local murder, as well as Africa's starving children, if the murder story does not fit some angle that the TV news editors are emphasizing or downplaying. For example, they may downplay a local murder story that does not supply evidence of increasing drug abuse or one that tends to upset racial harmony.

Moreover, the audience itself exerts its own selectivity. Television producers have to engage and keep a large audience, otherwise they cannot stay in business. This means that they have to select and arrange materials in such a way that they keep attracting their audience. Since they cannot long afford to have their viewers switch channels, they avoid challenging them unduly, badgering them, and, above all, boring them.

They may not broadcast too much harsh reality lest their Nielsen ratings drop and their ad accounts plummet. News teams know their highly competitive business must be both informative and also entertaining.

The demand for providing audiences with interesting bits of different interpretations of news events also imposes an excruciating selectivity on certain experts interviewed. Experts are people who have critically refined and nuanced their views within their specialized area of competence. For a news broadcast, experts in a particular area may be interviewed and filmed for a half-hour or more. Yet what they find eventually broadcast is often some ten-second remark of theirs, clipped out of context and counterposed to an equally truncated remark by a local expert of an opposing view. In this way, by opposing the views of experts to excite audience interest, editors have manipulated both the experts and the audience.

Of course, all the intervening cutting and selecting has usually occurred without the consent of either interviewee. Often neither of the opposing experts would have permitted such unbalanced presentations of their own views had they been allowed some voice in the actual editing of the news clip. Such interviewed experts are sometimes the innocent victims of media selectors who prefer to keep an audience attentive by giving equal time to all sides of an issue rather than to portray fairly their interviewed experts.

In addition, competent and well-informed experts cannot accurately communicate their views within the ten or twenty seconds which TV will allow them. Moreover, unlike popular TV personalities, experts may lack charisma and are not perceived as familiar faces that attract people immediately. So, the experts and their views will often be side-tracked while spokespersons from the government or corporations will be given the main track. These latter are eager to spread their own point of view, have been trained to meet the heavy demands of TV news clips, and will broadcast their brief report--carefully prepared, effectively delivered, and appropriately slanted.

The intent to attract and retain an audience also keeps TV personnel from dealing with serious questions, particularly in debates between political candidates. Here image production is the primary rule. Serious talk must be avoided; otherwise the program will not be entertaining. For above all, TV newscasts and debates must be entertaining and interesting enough for their majority of viewers with sixth grade mentalities. Otherwise, many of these folks will switch to another channel, Nielsen ratings will drop, and ad accounts be withdrawn.

Vice-president George Bush once showed his TV mastery at arousing audience interest. Pressed by interviewer Dan Rather to reveal his relation to the Iran-Contra affair, Bush used his political skill. Promptly charging Dan Rather with badgering him, Bush seized the offensive. He knew that to start a verbal fight on the screen would arouse the attention of his TV audience. On seeing him attack, they no longer cared about the Iran-Contra question but now wanted to see who would win the fight. In this way he dodged Rather's genuine question and distracted his viewers from the real issue. Bush thus increased his approval in the popular polls. In fact, many came to believe erroneously that, in serving the public, news personnel like Rather should not dig so hard for the truth, but instead should show more respect for the Vice President of the United States.

The need to lead an audience from the familiar to the novel is another factor that inserts much selectivity into media. If editors want to retain their audience, they may not start by flooding them with opinions clearly contrary to the audience's predominant beliefs and values. David Altheide put this directive simply, "[A]ny effort to present information that is not easily assimilated by a viewing audience committed to certain points of view will fail."[9] Yet to follow this rule is to inject an echo effect into the interaction between audience and media director. For then the director must at the start largely echo back to his target audience what they are familiar with and can assimilate. This puts tight limits upon the media director.

A still further constraint upon media transactions is the psychic associations that media audiences have already acquired. These associations call media editors to be selective in a new way. For instance, if the topic of Australia emerges, a good editor has to downplay effectively the emerging kangaroo image and the questions associated with it: "Have you seen one?" "Do their young stay in their pouches when they jump?"

Again, if Alcatraz is the topic, the editor needs to adapt, not to the associations of those sighting that island in the 19th century, but to those of the present. Prior to the 1849 gold rush, native Americans regarded it as their ancestral property. In the 1860's, ship captains viewed that small island as a landmark for steering safely into San Francisco Bay. Today's viewers of Alcatraz have quite different associations of that island than its nineteenth century viewers had. Thanks to previous media contacts with Alcatraz, people who focus on it today will, on the first mention of Alcatraz, think of convicts daring to break out. For today's

Americans have seen movies about "the Rock" and about escapes from its prison walls. Their first questions will be "Who escaped from Alcatraz?" "And how did he do it?' So, the editor of a story on Alcatraz must meet this mind set, which clearly rules out certain other kinds of questions. To be effective, then, an editor must take into account the psychic associations at work currently in the minds of his or her audience.

An experiment can help one to experience certain associations largely acquired through exposure to media. Contrast your earliest perceptions of an object over against your much later perception of the same object after TV advertising has mediated it to you. For instance, recall your first hand-me-down pair of tennis shoes and then the ads for a pair of clean, springy Reebok shoes. Then remember the grass in some rough and tumble playground of your youth over against the breathtaking view of a professionally manicured lawn as seen on TV. Contrast your first ride in a little old rowboat with that glamorous Caribbean pleasure cruiser advertised on TV. This mental experiment can alert people to the changes that TV has worked upon their consciousness. It may also alert them to the way people's interests and questions now selectively favor those objects that have ad appeal rather than those plain old things.

Moreover, by stimulating interest in certain topics and not in others, TV has worked great selectivity within its viewers. For instance, contrast Americans' sensitivity, first in the 60's and then in the late 90's, towards AIDS, cocaine, cholesterol, and oil slicks. Again, contrast how, first in the 60's and then in the late 90's, Americans valued racial equality, busing, unions, and saving money for future needs. Such examples highlight people's need to notice carefully still another way in which the media have influenced them. The media have selectively stimulated and fostered certain interests, starved out others, and in some cases over-saturated long-standing interests. For example, reports of the dangers of cigarette smoking have become so common that people are no longer impressed.

Furthermore, newscasts cannot visually present the invisible world. The subtleties of thought that link the circumspect discussions of expert analysts are often too specialized to be broadcast to average Americans who become vexed by thoughts they cannot grasp. Some TV editors may even foresee the profound long-range effects of the consultations and decision- making of corporate boards of directors. Yet they realize that these latter are also beyond the grasp of a popular audience.

Then too, remote, hardly accessible areas select themselves out of the media. Suppose that wire services report fierce battles in the Kashmir or in the Sudan, with casualties in the tens of thousands. The difficulty of getting camera crews to these remote scenes of action, plus Western audiences' slight interest in these areas, rule out such items from TV reports. Sometimes, too, a central media center (such as London, New York, or Los Angeles) may not receive a TV film report from such distant fields until after it ceases to be news.

Still other factors make TV news even more selective. Just where is the camera placed? Its position determines what it takes into view and what it does not. For instance, a camera focused upon a Washington reporter using the White House as a backdrop cannot simultaneously catch the nearby homeless beggars wrapped in newspapers and sleeping on the warm air grates along Pennsylvania Avenue.

Then too, how much more impact do visuals have over verbal materials? News editors have consistently shown Sadam Hussein looking mean and fierce and pictured Libya's Muammar Gaddafi raising his fist in front of armored tanks. In contrast, during the Gulf War they showed General Schwartzennegger smiling and dialoguing affably. Soundlessly, such portrayals carry significant political and ideological messages.[10]

Look at Leslie Stahl, anchor-woman for CBS, as she once broadcast a strong verbal critique of Ronald Reagan. The visual background for her remarks showed Reagan warmly embracing children in front of red, white, and blue balloons cascading freely to the ground. The strong visual impact of the winsome Reagan wrapped in patriotic symbols so overwhelmed Stahl's audience that they did not even hear her verbal criticism of him. In fact, the Republican Committee actually phoned to thank her for the free pro-Reagan publicity that her program had unintentionally provided.

Many forget that, like Nathan Rothchild's signal-sending system and Marconi's telegraph, TV is basically a product of capitalism. Hence viewers need to realize that at bottom TV has to promote a culture of consumerism. If it is dedicated to spread news, that spreading is first of all for those with wits and money enough to seize from it new opportunities for investing and buying. As political economist David Harvey put it:

> Access to and control over information, complete with a strong capacity for instant data analysis, have become essential to the centralized co-ordination of far-flung corporate interests. The

> capacity for instantaneous response to changes in exchange rates, fashions and tastes, and moves by competitors is more essential to corporate survival than it ever was under Fordism [that steady system of assembly-line production reigning from 1901 to 1972].[11]

To grasp the primacy of economics in telecommunications is to recognize the most dominant source of the media's selectivity.

News Unbalanced by Slants and Omissions

As we have seen, a news editor cannot avoid selecting news items. While some selection is unavoidable, slanted selection can and should be avoided. This latter differs from that other kind of selection of facts which intends to stimulate fair public discussion of genuine issues. Hereafter in this chapter, however, the term "selection" will be used as shorthand for slanted selection--that is, such selection of facts as intends to support unfairly one rather than another interpretation of the facts. Because publishers depend financially upon the ad accounts of corporations, slanted selection has favored and will continue to favor the interests of corporate America, not those of America's economically disadvantaged majority. As the cliché has it, clearly no publisher wants "to kill the goose that lays the golden egg."

Since controversial issues are most likely to tip the balance in media reporting, it may help to cite a study about fairness and balance of newspaper handling of these issues. Two researchers published their examination of the unbalanced way in which 18 selected newspapers covered controversial issues in early 1991. Their findings are here summarized:

> This study developed a content-based technique of assessing fairness and balance of newspaper coverage of controversial issues and applied it to 250 stories on 18 issues appearing in 18 newspapers during February 1991. The study found that, on average, one side dominated at least three of the six story qualities assessed. Only 7% of the stories were evenly balanced, while more that 50% of the stories had at least four of the six story qualities dominated by one side. Three issues involving abortion and two involving local schools were the most unbalanced. The Persian Gulf debate was the most unbalanced national issue.[12]

Such findings lead one to ask whether these instances of unbalanced reporting were intended or not, whether they stemmed from entrepreneurial bias, slants, omissions, or other factors. These findings bring to clear focus the pressing problem of fairness and balance in media work. And persons experienced in media work recognize how difficult it is to be as fair and balanced as possible. Yet that ideal identifies the genuine newsperson and serves as the norm that reveals unbalance through either slants or omissions or both.

Sometimes reporters and staff writers can omit to disclose their own perspective or how their own social position may affect their accounts. For instance, the *New York Times* covered the Anita Hill vs Clarence Thomas case with 80 news stories and news analyses. In them the writers assumed full objectivity and omitted confessing their own social identity as possibly affecting their writing. Instead, they suggested that everyone else involved in this case--Hill and Thomas, the senators on the panel, expert witnesses and citizens--all responded according to his or her gender, race or experience with sexual harassment Yet the reporters and staff writers never suggested that their unmentioned social identity may have affected their stories and analyses.[13]

An editor, too, can often skew a news story one way or another through omission. He or she may simply be ignorant of certain facts, or reject them impetuously, or even overlook them deliberately even though others have called these facts to the editor's attention. Omitting certain aspects of a story can keep people from thinking upsetting thoughts or raising troubling questions. Granted, at times omissions are beyond the control of editors and reporters. For example, reporters were denied entry to Grenada for many hours after the invasion began. Even more repressively, the U. S. Army permitted the broadcasting of only highly censored accounts of the happenings of Desert Storm.

Allowing for these restrictions, newscasts of Desert Storm suffered from still other omissions which media editors did create. As usual, they framed their news-reports as episodes rather than as themes to be analyzed.[14] In this way editors induced viewers of their newscasts to blame the people shown in the episodes for the sorry events portrayed. Had these events been treated thematically, viewers would have faulted the political or corporate leaders whose decisions created the events. As experiments have also shown, especially for TV newscasts, when editors frame news predominantly in episodes rather than in the themes underlying them, they favor the status quo of political and corporate incumbents.[15]

Moreover, in their analysis of news from Desert Storm, editors omitted to deal fully with a racial issue that clearly surfaced there in the U.S. Army. Try to see it, for instance, through the eyes of Dave Johnson, a middle-aged Black American, employed at a Boeing plant. Dave returned from work daily in Spring 1991, during and after the war in Kuwait and Iraq. On arriving home, he usually flipped on his TV set to watch the news. Dave is very intelligent, more perceptive than most TV viewers. What he saw puzzled him. During the war, the news scenes of U.S. ground forces could not hide the fact that about one-third were Black--more than twice the national average. Sometimes commentators even called attention to this fact but Dave noticed that they talked far more frequently about accurate bombings and successful technology. His puzzlement arose from what he thought was omitted.

Dave watched the generals and President Bush congratulating our troops and touting the bravery and outstanding performance of all units. This made Dave feel good about the Black men and women who had shouldered such a disproportionate amount of the fighting. Yet Dave found the media avoiding the root of the problem: why was the percentage of Blacks in the army more than twice that found in civilian life?

Of course, the media pointed out that young Blacks wanted a well-paying job and a good inexpensive education. Dave knew, however, that young, economically strapped whites wanted the same things yet didn't volunteer for the army nearly so frequently. Dave needed no books or lectures to learn how deeply racist most Americans feel, think, and behave. Their unconscious racism is, as he well knew, the more oppressive kind. Though not personally intentional, unconscious racism oppresses Blacks more since it is built into American culture. Dave found the media are turning attention away from, perhaps even hiding, the racist structures in American businesses, schools, and churches. These systemic injustices keep telling young Black males that since the doors of upward mobility are usually closed for them, they had better look somewhere else--like the army.

Dave remembered how the 60's had taught whites to avoid being overt about racist practices. As a result, they had merely learned how to be more subtle and effective in their racist attitudes. So, in the days after Desert Storm the news programs were subtly covering over a sore which media directors were reluctant to broadcast about either themselves or their fellow corporations or white American society. Dave may not have used the phrase, "selection of the news," yet he knew that TV editors were keeping Americans from facing up to their racist attitudes. He also saw how the media used their episodic techniques to keep ordinary viewers contented, even when the facts suggested that they personally should take effective steps to bring the dream of Rev. Martin Luther King, Jr., closer to realization.

Back in the 70's, the media hammered home the fact that of the approximately 2,600,000 U.S. soldiers who served in Vietnam, about 54,000 were killed there. The media omitted to mention that in the same war Americans killed and wounded about 6,500,000 Vietnamese. This omission kept Americans from asking such questions as: Are non-American people less important than Americans? Was so much killing proportionate to our losses? Why do non-Americans regard Americans as ugly? Yet Americans should honestly face questions like these if they really want to avoid dodging that healthy contact with reality needed for sanity.

For another instance, take the notorious omissions connected with reporting on Nicaragua in the 80's. After the U.S. Administration decided not to support the Sandinista Revolution of 1979, most American media did not publish well-documented reports that in the early 80's the Sandinistas were carrying out outstanding improvements in literacy, health care, schooling, feminism, and popular culture. Likewise, in their broadcasts and releases about the Contra war in Nicaragua, the U.S. media usually did not report the fact that the so-called "Freedom Fighters" (led mostly by Somoza's brutal National Guardsmen) eventually killed 30,000 Nicaraguans--mostly non-combatants.

Even in the 90's, seemingly trivial omissions still work their prejudicial effects. For instance, take two scenes of Black youth in a local newspaper. The one shows three American boys playing in a slum street, the other shows three African refugee lads trudging back home to Rwanda. Notice how under the picture of the Americans, the names of the three boys are given: Ron, Dion and Nikki. They are real persons who have hopes and fears, disappointments and successes. And although news photographers in Africa do get the names of the lads they photo, editors of American news media are likely to feel that such information is unneeded, or beside the point, or at least uninteresting to most of their readers. The result is that these African boys, named Paul, Didace and Innocent, "remain anonymous, a cipher. And it is easier not to care about a cipher."[16] Such boys are portrayed as objects. By this unfair treatment, readers are protected from caring and the prejudice is further inculcated that Americans are persons who count, non-Americans are objects who don't.

Turning to clearly intentional ways in which the media select the news, we can start with the choice of image that suits a publisher's purpose. Editors build upon the old Chinese proverb that one picture is worth a thousand words. When editing newspapers, magazines, and especially TV news, they skillfully select from tens or hundreds of images of a particular news story just that picture which fits in with and reinforces the editorial position. By a carefully selected image they call forth from their audience the response they wish.

For instance, when Gorbachev was in the limelight, what kind of response to him did editors desire? They chose from a range of photos showing Gorbey smiling and confident, through a whole gallery of shots, down to those picturing him as worried and ashamed. It was for the editor to choose but his choice tipped thousands of minds. Again, editors can make Bill Clinton seem a caring human person or a fearful, indecisive leader. An editor simply has to show the president smiling warmly at a little girl while paternally shaking her hand or to flash a picture of Clinton standing with worried face and wrinkled brow as news-reporters press him hard with questions.

Rather than selecting one image, TV editors and designers of commercials have another technique. They can so rapidly shoot out a whole series of varied images that viewers overlook the one realistic image they need. For TV's sweeping flood of ever varied images trains the minds of viewers to favor change of any kind rather than to stay steadfast and be consistent. Television's passing current of images also tends to erode our vital evaluation of those critical historical events that we hold in common--for instance, those identity-shaping events of our English-speaking nation. Television can flood an audience with so many clips of the past that viewers' minds are as equally unimpressed by the signing of the Magna Carta in 1215 as they are with the a scene of Englishmen eating porridge in that year.

In this way historical documentaries on TV sometimes juxtapose items while omitting to highlight the significance of the historically key event. For instance, they may link pictures of Columbus' setting foot on the New World with Queen Isabella's fine-laced sleeve, without signaling to an audience of sixth-grade mentality which is the more important of the two. They may associate the Boston Tea Party with some new crockery used at Williamsburg in 1776. They may link the battle of Gettysburg with Grant's muddied coat at Appamatox. Finally, they may show the Japanese attack on Pearl Harbor with the drinking of a particular cocktail preferred by upperclass Hawaiians in Fall, 1941.

If editors fashion non-evaluative associations this way by omitting needed highlighting, it tends to erase Americans' appreciation of our nation's identity-shaping events and our consciously shared attention to those crucial moments in our national history. In brief, since TV's stream of images tends to flatten out its contents to an equality of being merely next to each other, these de-valued images can numb us historically and thus erode our sense of identity as Americans.

Sourcing

Another usually intentional way of selecting news items lies in sourcing. United States media are highly selective of the sources upon which they base their news broadcasts. Yet most of their audience do not suspect this slant. Because of its increasingly tight homogenization, the U.S. news sources are limited practically to two wire services: the Associated Press (AP) and the United Press International (UPI) . Other wire services are generally not employed, although the recently founded, solution-stressing American News Service (ANS) may insert a fresh approach. Currently, the UPI, as a for-profit organization, has been undergoing financial problems and consequent downsizing. The AP, founded in 1841, is the world's largest news-gathering agency. Worldwide it employs 3,400 personnel--some full-time, some *ad hoc* assignment reporters. As a non-profit cooperative, AP is owned by media networks, newspapers, TV stations, and other news publishers. As a news-gathering agency, AP is fed by reports from its owners and in turn supplies them with its edited news. The local, state and national offices of AP sift through all incoming news stories, edit them, and then depending on their appeal and import, send them out on wire networks for state, or national, or world-wide distribution. How the Internet--for example, with its American News Service--will impact on AP and UPI in the near future remains to be seen.

Appraising news-gathering from trouble spots on the international scene, Francis D. Gomez, a government observer at Georgetown University's seminar on the handling of news from conflictive areas, put it this way, "Most American papers rely on U.S. wire services. Excellent items by Reuters, AFP, EFE, ANSA, and other agencies are missed by the American public, the Congress, and often the Executive."[17] This limited sourcing quite effectively excludes the voices of billions in the earth's global village.

This limitation is intensified by the excessive influence what the U.S. Government and large corporations exercise upon the AP and UPI. In this way, as Chapter 7 will show, U.S. government officials have the capacity to flood the media with distorted information that effectively sets the parameters for debating crucial issues.[18] Sometimes 60-70% of the new releases from AP or UPI originate from the U.S. Administration's flooding of these news services. To some extent this flooding resembles the technique of TASS, as Chapter 7 will also show. Americans demand a great variety of sources to obtain their food and clothing. Yet, when it comes to news, they surrender to being fed just by AP and UPI for the most part. These two news services preponderantly supply the three major TV networks and such major newspaper chains as Gannett, Hearst, Scripps-Howard, Time Warner, and Knight-Ridder.

A noteworthy instance of this occurred in the case of the withdrawal of the U.S. from UNESCO.[19] How did the AP choose its sources during its four month coverage of this important debate? The AP allowed only 3.6% of its soundings to be taken from that two-thirds of the world's population to whose development UNESCO is largely directed. The AP compressed all the views of ordinary Americans, of UNESCO project supervisors, and of professional political scientists at home and abroad into only 7.3% of the whole spectrum it used as its sources. From UNESCO officials who almost unanimously opposed U.S. withdrawal from UNESCO, AP derived only 22.2% of its material. From officials in the governments of our Western allies, who were far more divided about withdrawal, AP derived a further 18.1% of its data. But from U.S. government sources, which almost unanimously favored withdrawal from UNESCO, the AP drew a walloping 43.1% of its data. For the AP to grant only a 7.3% sampling to the foreign and American political science experts and then an incredibly low 3.6% sampling to that two-thirds of the earth's population mainly served by UNESCO seems at least a power play.

Sourcing like this also manipulates the minds both of members of Congress and of most Americans. Neither of these two groups have a voice in saying whose opinions shall be consulted, or how much they should be valued. The U.S. Administration had already decided to withdraw from UNESCO and the AP was simply trying to produce in most Americans a general agreement with that decision. Not surprisingly, the rest of the so-called elite free press--*New York Times*, *Washington Post*, *Time*, *Newsweek*, and *U.S. News & World Report*--all marched in close conformity with the Associated Press.

Priming

Still another direct way to select news items lies in priming. Here editorial boards instruct reporters to ask certain questions of administration officials or political candidates. By thus taking the leading step (or priming), news reporters get the others to dance accordingly. David D. Newsom, of the Institute for the Study of Diplomacy, holds that a media spokesperson needs guidance for the probing questions he or she shall put to the Administration at its noonday briefing session. Newsom goes on to assert that media directors' selection of these key questions "probably creates more U.S. foreign policy in a shorter space of time than any other part of the process [of creating foreign policy]."[20]

When editors give problem X a prominent position while omitting or giving short shrift to problems Y and Z, they prime the minds of their audience to regard problem X as far more pressing than Y or Z. For example, in pre-election days when press agents ask political candidates certain questions and not others, these selected issues become the criteria for weighing the performance or policy of the candidates involved.[21] Sometimes, a spontaneous priming occurs, as, for instance, in the second presidential debate between President Clinton and Senator Robert Dole at San Diego in October, 1995. Only one of the approximately two dozen randomly chosen sample of audience questioners asked about foreign policy and none inquired about abortion. How balanced and fair is such random priming? This question might wisely have been pondered by American viewers of the debate before they cast their vote.

News editors and writers are familiar with the wider range as well as the limits of their craft. Generally they attempt to maintain a posture of neutrality towards public figures. To this end they employ their diversity technique which counterbalances pro and con views. They also broadcast events that both enhance and diminish popular approval of a politician.[22]

Nevertheless, without resorting to either advocacy or attack of a politician or of his or her position, and without departing too far from objectivity, media personnel can bend the truth rather than break it. If they nuance a phrase or place emphasis skillfully, they can convey their own position along with the news. This lack of balance in the news reveals itself concretely in certain kinds of lead-ins or headlines, the use of certain photographs (attractive or repulsive, as needed), and the choice in TV and radio of various kinds of music to accompany different news stories.[23] By innuendo, association, comparison and contrast, as well as by other embellishments, news editors can create the impression they desire.

Some newspapers are so tied into the chain that owns them--like Gannett or Scripps-Howard--that their editors stay largely within the lines set by directors of the chain. Certain papers may be so attached to a political party that they are unofficially its local mouthpiece. As a result a particular city like San Francisco or Cincinnati can be blanketed day after day with conservative thinking. By contrast, New York and Washington can be blanketed with liberal thinking. The editors of the *New York Times* and the *Washington Post* are fully aware how attentively members of Congress scrutinize their pages before going to their offices each day. Moreover, these editors know that what makes the front page of the *Times* or *Post* often sets the tone of questions that reporters will fire at Administration officials in upcoming news conferences. Other news items may affect the common good of most Americans more urgently but these editors control their front pages.

Some editors use downplaying and trivializing to control thought. Sometimes maintaining public peace and security requires the downplaying of rapes and interracial murders. In other cases, however, the tilting hand of an editor shows itself. In 1996, the Pentagon released documents confirming critics' claim that the School of the Americas at Fort Benning, Georgia, had for decades been teaching blackmail, torture, and execution to 60,000 Latin American officers and deserved the nickname, "School of Assassins." On November 16, 1996, the anniversary of the brutal murders of six Jesuit priests and their two companions by Salvadoran army officers trained at the School of the Americas, 400 people gathered outside the School to protest the training of terrorist soldiers for Latin America. Fr. Paul Locatelli, Jesuit President of Santa Clara University, publicly called for the closing of the School. When 60 cross-carrying protestors stepped onto the School's grounds, they were arrested. Major news sources and the elite press played down this event.[24]

During the Gulf War, major protests in many cities across the country were barely noticed on national TV, although a few hundred pro-Bush flag-wavers in Columbus, Ohio, were given extended coverage. In 1995 Pope John Paul addressed the U.N. and called its members to care for immigrants and the destitute of the world. Finding this message too challenging for his audience, the national news editor of one TV network chose not to give this papal story the lead-off spot it clearly deserved. Rather, he assigned it to fifth place, coming after TV scenes of a local fire in California and a new car show in Vietnam.

Minimizing or exaggerating the number of participants in a gathering also unbalances the news. 207 people actually protested outside a local Federal Building, reporting that about 50 people did so keeps general interest effectively unarroused. Was not the 1996 national march on the Washington Mall by Black Men United first reported at about 400,000, when the actual figure came closer to one million?

The stereotype, however, remains a most powerful way of presenting news in an unbalanced way. To picture smokers as exuberantly healthy conveys a misleading signal. To display honest people unfortunately unemployed as sitting shiftlessly along sidewalk benches reinforces associations of "lazy ne'er-do-wells." To focus on the glitz and glamour of the few who belong to the financially successful class while completely omitting the hard work and struggles of the laboring class is to unbalance the view of that reality which is America.

Giving women the silent treatment stereotypes them as no-accounts. In England, at the death of Labor Party leader, John Smith, in May 1994, Margaret Beckett (as deputy) became interim leader of the Party during a difficult period of transition. Yet when the question arose of a proper Party leader, she was immediately dismissed as a viable candidate. When the English press from the start tilted toward a male candidate, it revealed how press patriarchy works in practice and how the boys' networks operate concretely.[25]

In America a different kind of sexist stereotype is found in TV. As pointed out in *Big World, Small Screen*, the American Psychological Association's 1991 report on TV:

> [A]lmost one quarter of the women shown on television in 1984 were extremely affluent, whereas in reality only 0.2% of women earned over $75,000 a year. A much higher proportion of working women are shown in professional and entrepreneurial roles than actually exist. Moreover, despite the increased portrayal of working women, [on TV] these women rarely experience problems with child care, sex discrimination, harassment, or poverty.[26]

The same report also found that white male adults get the lion's share of TV exposure but that women, children, ethnic minorities, and the elderly are: ". . . under-represented on television relative to their numbers in the population. For most, television also conveys negative stereotypes. Women and elderly people are often portrayed as weak, inadequate, and incompetent."[27]

Actually, the daily grind of working class people is rarely mentioned in schools or portrayed on TV screens. The International Association of Machinists and Aerospace Workers did a study of how working people were portrayed in the media. This 1980 study uncovered the following stereotypical attitudes toward America's ordinary work force: under-represented, patronized, dumber, more foolish, and less competent, even though more friendly, funny, and less selfish than white collar workers. Working people were also portrayed as simplistic, "serve and obey" personalities who drank and smoked more than upper middle-class people, and were also less attractive. No hint was conveyed that these workers lived under the ever present thumb of their employers and suffered the dire threat of being fired summarily if the "bottom line" could not support them.[28]

All these ways of selecting and unbalancing the news are distasteful to notice. Yet they increase the evidence supporting Scott Peck's designation of many Americans as people caught in the lie. Even though we are indebted to the media for its many kinds of often outstanding service, the call to be watchful of the media and to protest critically against their methods of misinformation becomes increasingly urgent.

By itself a proper balance and selection of the news does not solve the problem of achieving fair news coverage. For inaccuracy and obscurity might still mar the news broadcast. These concerns occupy the next chapter.

QUESTIONS FOR DISCUSSION

1. What positive traits can you find in the way the news is presented?

2. From recent newscasts give examples that indicate the media's dominant economic interest.

3. What are some unavoidable constraints on reporters and editors which result in selecting news items?

4. How can editors intentionally select the news of the day?

5. Give some instances of government intervention leading to slanted news?

6. How does the selection of a particular image give an editor power over his audience?

7. Why do you think most news media depend on only two sources--AP and UPI? Where do these wire services get their information?

8. Give some examples of the way stereotyping unbalances the news?

Chapter 5

ACCURACY AND CLARITY OF THE NEWS

At the beginning of the 20th. century, forces began propelling the presentation of news in a different direction. Eventually, those forces became far-reaching and all-pervasive. For until the mid-1900's, editors felt an obligation to comment on the news in order to provide an understanding for their readers who were not well educated. For as factories became more mechanized and able to produce more goods that cost less, journalism began changing. Advertising and marketing agents demanded a wider circulation of newspapers in order to reach prospective buyers for these more available goods. So, in their attempts to increase circulation, editors stuck as closely as possible to a neutral rendition of the news so as not to offend readers. Big cities with a diverse population called for facts and bulletins rather than editorials or commentaries. Today, less space is given to news and more stress is placed on ads and local human-interest material.

Influence of Theodore Roosevelt on Journalism

To understand changes in the communication of news, we need to consider how Theodore Roosevelt influenced journalism. Seeing how increased circulation affected editorial policies, Roosevelt was the first president to exploit the situation and set an example for future presidents. He realized that news and its presentation molds public opinion more than editorials. To exploit this shift, he needed to seize the headlines and influence the way reporters wrote about him. He kept reporters supplied with news guaranteed to dominate the front page. He paid close attention to the way the news was disseminated. A veteran reporter remarked, "President [Teddy] Roosevelt knew the value and potent influence of a news paragraph written and disseminated through the proper influential channels better than any man who ever occupied the White House, before him or since."[1]

Many years later, a reporter recalled the changes that took place in the relationship of the president and press when Theodore Roosevelt became president. He wrote, "Roosevelt let us into the White House--the first time in American history. He established the White House press corps and gave us a room to use in the White House--one way to get us on his side."[2]

Roosevelt made sure that the press was nearby to make it easier to generate news about himself. He used the "trial balloon" as one of his pet devices. In order to test how the public would react to a proposed bill, he leaked a bit of information to a reporter. If the public reacted positively, the proposal became policy and the reporter received praise for his fine work. If the public responded negatively, the information was denied as misinformation and the reporter denounced as incompetent or called a liar.

Once Roosevelt used the technique of blanketing the news, that is, denying publicity to others. Charles Hughes was scheduled to announce his candidacy for president and explain his views on national issues in a speech in New York on the evening of January 31, 1908. Roosevelt made sure that Hughes received very little publicity. He leaked to reporters that he would deliver an important message to Congress on January 31. They waited all day as the deadline for the evening papers came and went. Then, at 6:00 p.m., the President appeared before Congress and asked for long-awaited legislation making workmen's compensation available to federal employees. It is not hard to imagine who got the headlines the next morning.[3]

Influence of Journalism on Politics

Following the example of Roosevelt, presidents, government officials, and politicians find it greatly to their advantage to get favorable publicity, to have their views and programs brought to the attention of the widest possible audience. Many people have little time or inclination to get more than superficial information about the news. Tired after a day's work, they prefer to spend the evening watching sports or movies on TV. They tend to believe that the press is out to get the president. Media psychologists, Grossam and Kerman conducted a two-year study to test this ordinary perception. They concluded that the cooperative elements of the relationship between the president and press are at least as strong as the antagonistic elements. As an instance, they pointed to the 1968 elections when the press certainly knew what Nixon was like. Nevertheless, 499 daily papers endorsed him while only 93 endorsed Humphrey.[4]

Before TV, politicians depended heavily on party support during their campaigns. Now, according to Hal Evry, "If you appear to be all things to all people, get on television and don't say anything but make it sound good, you can get three out of four to like what they read into you."[5] Television has also taken the place of newspapers in determining the outcome of presidential elections. As James Deakin puts it, TV so easily distorts the truth "with its winks and sneers and lifted eyebrows, its carefully edited film footage and its cleverly slanted 'commentaries' by Ivy League liberals directed to an audience of millions whose intellects are presumed to be despicable."[6]

Today, computers are used to match particular audiences with specific political issues. In depth studies are made into the life-styles, hopes and fears, fantasies, personality traits, emotional needs, prejudices, and ideas of a group in relation to a candidate's behaviour and appearance. This process of psychographics has been so perfected that politicians can target groups as specific as individual neighborhoods.[7]

Reporters experience many pressures from political incumbents and hopeful campaigners. Yet if the news media do their job well, they draw attention to mistakes, abuses, shortcomings, and corruptions of those in power. They do this impartially regardless of which party or individual is in office.[8]

Goethals, however, maintains that for a limited period of time each day, our attention is turned to

> . . . highly concentrated symbolic reports of events. These thirty-minute distillations, unlike detailed, lengthy explanations, give a sense of the totality of things--a selective, fabricated view of "reality." What we actually view are shared symbols of the real world. We may know very little about what is happening in, say, Nicaragua, the Middle East, or Washington, D.C., but we grasp certain symbolic representations which become the basis of thinking, acting, and voting. Without prolonged reading or investigation, constant viewers of television are indoctrinated through words and pictures into a world view. When TV's words and pictures are our only source of information about the lives of people in distant countries, their land, government, and living conditions, the

> stakes are high. In a free society, a great responsibility rests on those who produce news programs.[9]

Distortions in the News

As people become more aware of some distortions in the news, they begin to ask over and over, "Why don't the news media tell the "truth". They argue and argue about this question with their friends, little realizing the futility of the question.

The complexity of today's world and the limitations of time and space, force the media to use other standards of "truth," such as accuracy, fairness, and completeness. To add to the confusion, in journalism there are at least three kinds of accuracy:

1) Factual accuracy--reporting of distinct facts;
2) Contextual accuracy--putting the facts, ideas and events into their historical perspective;
3) Linguistic accuracy--choosing appropriate words for the report. The way something is said may be more important than what is said.[10]

The amount of space allotted to actual news in the daily paper is pitifully small, and the evening TV news must share the half-hour with commercials, sports reports that include replays, and human interest stories. It follows that incompleteness becomes a cardinal flaw of journalism and a source of distortions which are, at times, almost impossible to avoid.

Also, the complexity of any new legislation proposed by Congress or the President dooms it to distortion in the news. Copies of it given to reporters may run 20 single-spaced pages while its backup commentary fills 50-100 pages. At the White House press briefing, reporters fill pages in their notebooks. Faced with all this material, the editor is expected to produce a segment of 100-120 words for a TV spot to air on the evening news.

To further complicate the task, every major issue has background in the past which should be considered in the interest of accuracy and fairness. For important issues, committees and subcommittees have generated reports, and background studies that may report decades of previous experience. If interviewed, experts could throw light on special aspects of the issue. How much time do reporters have to do the in-depth research that would make their story complete and accurate?[11]

Each human being tends to believe that he or she has the facts and that therefore those holding a different view are wrong. James Deakin points out some of the difficulties journalists face in presenting facts:

> The journalist defines his professionalism as the pursuit of facts and explanations. Facts are uncomfortable things: they embarrass officials and institutions and organizations, nag at ordinary people, disturb the status quo, challenge accepted practices, affront complacency. The public says they want only facts, but facts demand explanations and point to conclusions. Explanations and conclusions are controversial, and they open the way for bias unless various possibilities are given.[12]

Given these aspects of the news, some distortions seem inevitable and beyond the control of the media. The media, however, must bear responsibility for other distortions. By favoring certain groups--whites over blacks, affluent over poor, males over females, young over elderly--a distorted picture of society is constantly kept before the public. For instance, in a random sample of 100 prime-time TV shows, only 2.8% of their 1,228 speaking characters were judged to be over 65 years old.[13]

Often Ronald Reagan cleverly distorted the news by labelling. During his administration, news items about Nicaragua invariably mentioned the "Marxist leftist government" in contrast to the "Freedom Fighters" who were receiving arms and support from the United States. This attitude was in keeping with the philosophy of the Santa Fe Report, prepared by the Council for Inter-American Security in 1980. This report states, "Foreign policy is the instrument by which peoples seek to assure their survival in a hostile world. War, not peace, is the norm of international affairs."[14]

The endless repetition of certain words in the news distorts the reality unless consciously re-interpreted by the reader. The public has largely come to accept "defense spending" for military spending, "bailout" for the huge amounts of tax-payers' money paid to savings and loan companies to cover bad investments, "experts" for those promoting a favored point of view, "intelligence community" for the CIA and various spy agencies, "extremists, fanatics and pacifists" for groups protesting injustice. Intelligent reading of a news item in the daily paper becomes an exercise in translating media jargon into the language of reality.

United States diplomatic and psychological warfare techniques converged during efforts to discredit Nicaragua during its elections in 1984. A group of U.S. and international observers were present in

Nicaragua during the elections, trying by their very presence to encourage fairness and unhampered voting. After their return, their daily paper surprised them with stories about a reported MIG crisis in Nicaragua. According to disinformation sources from the United States, Nicaragua would soon receive advanced fighter jets from the Soviet Union. Every day these observers checked their paper for further news of the crisis. The U.S. reaction to the crisis appalled them. Democrats and Republicans in Congress vied with each other trying to justify direct U.S. military action when the MIG jets arrived. The United States cancelled leave for troops at Fort Bragg and mobilized thousands of others for land, sea, and air maneuvers off the coast of Nicaragua. Imagine the surprise of some Americans who heard from a friend in Managua that the United States had simply fabricated a "MIG crisis" in an effort to distract Americans and others from the positive news of Nicaragua's electoral process. In reality, the ship's cargo did not contain MIG jets but donations of toys for the upcoming Christmas.[15]

One frequently documented form of distortion arises from stressing the official position on an issue and downgrading popular grievances or protests. Such distortion was often used during the Persian Gulf War. Victorious exploits of the Coalition Forces and Bush's rising popularity rating filled the news while protests and demonstrations against the war received only passing and infrequent mention.

Even when such protests were mentioned, their importance was easily downgraded by reducing the number of protestors. How much publicity was given to the 35,000 who rallied against the Gulf War in downtown Seattle? How many people realize that on January 15, 1991, 15,000 people blockaded the Federal Building in San Francisco and shut down the building for six hours? At the same time, thousands marched to the Oakland Bay Bridge and blocked traffic there. Just two days later, another demonstration of 10,000 again closed the Federal Building while thousands marched to the Stock Exchange. On these two days in San Francisco alone, 1,600 were arrested. National news mentioned all these events only in passing, if at all.[16]

Language plays an important part in the distortion of news. During the air strike phase of the Gulf War, generals from the area reiterated that the Coalition Forces were "softening up" the enemy when they meant they were killing thousands of Iraqi troops and civilians. The tone of the briefing sessions held every day sounded more like the report of a football game than the life and death struggle of human beings. When not permitted to witness the action, reporters felt dissatisfied with the quality of the information supplied to them. The military took great care that the public not receive information that might cause a reaction like that against our activity in Vietnam. The military learned in Vietnam the necessity of the strictest control of information whenever public opinion can influence the political process. Before and during the Gulf War the images and interpretation of events presented by the media directly influenced public opinion in support of the military action.[17] For example, an effective propoganda story about Iraqi atrocities in Kuwait was the so-called eye-witness account before a Congressional Human Rights Caucus. A young Kuwaiti woman gave a dramatic account of Iraqui soldiers removing babies from incubators and leaving them to die. It later emerged that the witness was the daughter of the Kuwaiti ambassador to the United States. She had been coached by a public relations firm hired by the Kuwaiti Government.[18]

Film coverage of the bombing of retreating Iraqui forces was fundamental to the decision to end the war. It was feared that such images would adversely affect public sentiment if the war continued. Once the live phase was over, another strategy took over. The vast amount of information suddenly released, including taped audio-visual records, caused the "event to become utterly dispersed into a morass of conflicting interpretations and hypotheses about what really happened."[19]

By emphasizing the power of our technology, the military distorted the reality of our killing 100,000 Iraqi soldiers and the subsequent starving of 150,000 Iraqi children and non-combatants that resulted from the destruction of the infrastructure.

Suspicious sources and unattributed assertions frequently lead to distortions and sometimes even to false information in the news. Often reporters cannot divulge their sources because the people supplying the information may lose their jobs, be in danger of injury or even death. Whistleblowers in government or large corporations are especially at risk.

Unfortunately, journalists often give the same kind of anonymity to official sources who should be held responsible for their actions. Readers should beware of information that comes from an unnamed government official or agency lacking a legitimate reason for hiding the identity of the source. Usually such information consists of a "plant," inserted into the communication process to skew the news item. When phrases such as, "U.S. intelligence experts believe . . ." or "unnamed intelligence sources state . . ." or "CIA officials warned . . ." appear in the news, one wisely interprets them as warning flags of the possibility of misinformation or lies.

Suppression and Disinformation

The seriousness of suppression and disinformation in the news becomes evident from the White House's own worries about the adequacy of its intelligence-gathering system. In an article entitled, "Secret History of the War," *Newsweek* included a boxed insert of "Intelligence Goofs" as follows:

> The White House is seriously concerned about three major intelligence failures during the Gulf War:
> 1) U..S. planners were told the Iraqis had moved 540,000 troops into the Kuwait theater, with more than half in Kuwait itself. The actual total was nearer 250,000 of whom about 150,000 were in Kuwait.
> 2) Iraq was thought to have about 35 mobile Scud missile launchers. In fact they had as many as 200.
> 3) Intelligence reported Iraq had moved large numbers of chemical weapons into the Kuwait theater, and major efforts were made to destroy them. Not a single chemical weapon has been found [in Kuwait].
>
> What concerns policy makers is that these failures come on the heels of errors in East Europe. U.S.intelligence never picked up the Soviet Union's huge redeployment of weapons east of the Urals, so that they wouldn't count under arms treaties. And in East Germany, major Soviet installations, including a marshaling yard for missiles buried beneath a hill, went undetected.[20]

The way some stories are played down or suppressed severely flaws our information system. Also, our world view is significantly distorted not only by misformation but also by what we don't know. For example, South Africa succeeded in suppressing the powerful images that would have threatened its apartheid policy if the information had reached the West.[21]

Government control of commercial satellites could amount to unconstitutional prior restraint, a possibility that concerns news organizations. According to present regulations, the Secretary of Commerce with the Secretary of Defense have a certain amount of "shutter control" if national security or foreign policy interests are compromised. Mark Brender, ABC news producer and founder of the Remote Sensing Task Force, comments, "standards by which government decides to restrict should be very high so that government is not able to issue shutter control orders merely based on political whims, political senses, and frivolous concerns."[22]

If a broadcaster suppresses or distorts the news, the public interest suffers. First, misinformed citizens cannot participate effectively in the democratic process. Then, no true debate on public issues can be conducted if it rests upon a distorted presentation of the underlying facts. Many Americans became aware of information being suppressed during the Gulf War and felt unable to form an enlightened opinion or discuss the issues intelligently.

Real-time reporting has created tremendous problems both for the U.S. military and for the media. The military is concerned about the possibility of a reporter airing information detrimental to national security. Many in the media fear that military censorship based on real issues of national security might lead to political censorship based on political expediency and the desire to preserve the administration's image.[23]

Closely allied to suppression of news is news management by "disinformation." In October 1986, the press disclosed a memo to Ronald Reagan from John Poindexter, national security advisor, regarding a strategy that would assist the administration in its Libyan policy.

> In his memo to the President, Poindexter explained how it [the strategy] could be used against Libyan leader Muammar Gaddafi. The essential factor in disinformation strategy is the combination of real and illusory events to influence the public's perception of reality. Poindexter argued that the government had a right to use disinformation as a strategy to protect national security. Denying that it was intended to deceive the press or the public in the United States or elsewhere, he still conceded that some people might have been deceived.[24]

The proliferation of talk shows has added to the amount of disinformation presented on TV. Quotes from hardcover books that claim to have inside information about government and the lives of politicians fill the air waves with false information. Margaret Carlson accuses the press of catching the "Mad Lie Disease." Using the example of a book by Gary Aldrich, an FBI agent inside the White House, she writes:

> In his book, *Unlimited Access*, [Gary] Aldrich claims he helped decorate the White House Christmas tree with sex toys in 1994. This could not be true as all decorations are screened, logged and photographed in October. After Brinkley shot down some of his (Aldrich) statements, it gave new life to the book. The next day it sold 88,450 copies.[25]

Rigging and Slanting the News

The Federal Communications Commission (FCC) has warned against rigging or slanting the news by calling it a "most heinous act against the public interest; there is no act more harmful to the public's ability to handle its affairs."[26] The FCC has received serious complaints of news distortion in all four categories of rigging or slanting the news: selectivity, inaccuracy, misrepresentation, and staging. The hardest of all to substantiate, however, are complaints of misleading selectivity. Given the amount of material that reaches an average sized newspaper each day, selectivity of some sort is inevitable. It would be impossible to print or report all the news. It has been estimated that an editor examines 110,000 words of wire news daily, and 5,000 words from the local staff. This is equivalent to any average size book. Choices must be made to meet the deadline for publication. Usually five criteria are used to choose news stories:

1) Local news impresses local readers more than a traumatic report from a distance. A severe thunderstorm nearby has more impact than a serious earthquake in India.
2) Natural or man-made violence, conflict, disaster, or scandal get special attention.
3) Items pertaining to well-known people or familiar situations interest many people.
4) Events close to home interest many readers.
5) Novel happenings, especially if timely, catch the eye.[27]

These criteria for selectivity omit any consideration of the over-all significance of the news item. Often reporters or editors so personalize information of over-all significance--for example, an increase in unemployment--that the main point of the story becomes lost or trivialized. As mentioned in Chapter 4, the item is treated as an episode and not as a theme. By not treating the item as a theme calling for reflection, editors focus readers' minds away from the authorities responsible and onto the individual victims to be compassionated[28]

For example, Bill Mullins, a man in the Midwest, lost his job because his auto-assembly plant moved to Mexico. His picture with his wife and children, along with a long interview appeared in the local newspaper. The reporter asked Bill how the family would manage now as he has few marketable skills. The news item aimed at making the TV audience feel sorry for Bill. The program did not mention the deeper issues involved--the consequences of increasing unemployment or the ethical issues involved when large corporations tend to relocate for cheap labor.[29] Slanted news often results from emphasizing certain aspects of the story. For example, during the Gulf War, the media focused on the power of U.S. technology with little or no mention of the lives lost as a result of using that technology. As we have seen, the American public remained generally ignorant of the number of Iraqi children and non-combatants killed during the conflict.

Richard Schickel, commenting on the Rodney King riots in Los Angeles, shows how TV warps our reception of the news. He says:

> Television's mindless, endless (generally fruitless) search for the dramatic image--particularly on the worst night, Wednesday--created the impression that an entire city was about to fall into anarchy and go up in flames. What was needed instead was a geography lesson showing that rioting was confined to a relatively small portion of a vast metropolis and that violent incidents outside that area were random, not the beginning of a concentrated march to the sea via Rodeo Drive[30]

Reporters were trying to do their job but TV anchors kept asking questions designed to keep viewers frozen in fear as they watched in their living rooms.[31]

Richard Schickel continues:

> Their masters provided these journalists with almost no opportunity to do what many of them manifestly wanted to do: interrogate authority about strategy and timetables; question experts who knew something about the patterns of urban unrest; follow up a hundred human-interest stories."[32]

Editing, Cutting, and Staging

Also, misrepresentation can occur through editing techniques whenever a news item includes long interviews. By splicing answers given to a variety of questions, the editor can create a new and misleading answer to a single question.

Jeremy Murray-Brown describes another kind of editing technique, called the "cut," the shifting of the camera from one angle to another. He claims that the cut determines "the hidden message of the medium," for it gives a message to the viewer to look for meaning. Linking images in this way forms a webbing of pictures that has been artificially created. Goethals comments on Murray-Brown's analysis: "Since news stories are relatively short and must be dramatic, the cut becomes the organizing principle governing their pace and emphasis."[33]

Another editing technique, called "staging," involves a significant event which either did not occur or occurred at another time but is now acted out at the request of a news reporter. Staging is sometimes used to provide a dramatic effect to an otherwise dull event; for example, MacArthur's return to the Philippines was a landing staged well after the Marines had re-occupied the Philippines.[34]

Unintentional distortion of the news arises from the size of the United States. Up-to-the-minute stories from the West Coast seldom appear on the East Coast because they fail to meet the scheduled deadline for the evening news. In a national election, publishing early results in the East can have an adverse effect on the results in the West where the polls are still open. The problem has been highlighted for years at the time of presidential elections, but nothing has been done to change the situation.

In order to avoid any semblance of censorship, the FCC usually does not intervene in disputes over the truth of news programs. The FCC will investigate only when extrinsic evidence is provided. Extrinsic evidence is defined as testimony from "insiders" or persons who have direct personal knowledge of an intentional attempt to falsify the news.[35]

Imaging Less Developed Countries

Television has a major propaganda function in the area of news reporting. Its reporting of international (as well as domestic) news has always evoked complaints about misrepresentation. This issue, however, has recently received an unprecedented amount of attention because less developed countries charge that their images are distorted in the Western media. The General Conference of UNESCO in Nairobi in 1976, appointed the McBride Commission to study the "free flow" of information. Their report, *Many Voices, One World*, maintained that:

> [T]he image of the developing countries is frequently false and distorted. More serious still, according to some vigorous critics, it is this false image, harmful to their inner balance, which is presented

to the developing countries themselves.[36]

Unfortunately, many in the West have not accepted the concept of developing a New International Information Order which would provide more balanced coverage of news from and to developing countries. Media directors and editors still look mainly to economic profit rather than intending primarily to promote the genuine liberation of developing peoples. These media personnel ask how they can sell less developed countries still more access to satellites, TV programs, video and audio cassettes, wire services, computers, copiers, etc. However, they are unwilling to steadily mission bi-lingual reporters to less developed locales to do in-depth stories aimed at helping these indigenous peoples and also Western audiences appreciate the strong humanizing values of many non-Western cultures.

In news items from less developed, images of famine, corruption, and conflicts still predominate on our nightly news broadcasts. Thus powerful visual images are presented to TV viewers in broadcasts that usually lack time to develop the stories and provide adequate explanations. The "shorthand" nature of TV news lends itself to such distortion, thus creating a form of indirect propaganda that affects our perceptions and shapes our attitudes toward a wide variety of issues. Moreover, when received in developing countries, this distorted news often fixes these people further in depression and envy.[37]

In the process of economic development, any nation loses some part of its traditional culture. In their struggle for economic development during the 1960's and 1970's, many nations in Africa and Latin America accepted help from the most developed countries. Not only did this aid often impede economic progress but it also increased cultural dependence, largely through electronic media controlled by the U.S. and other Western countries.

Recognizing the trend, attempts were made in the 1980s to achieve economic development and at the same time prevent a disruption of cultures. Leaders in developing countries now realize that independence is necessary for all kinds of growth. As Anthony Smith sees it, the consequences of North-South relations will tend to work havoc on the developing nations. He says:

> Without independence in information and culture the gains of political and economic independence are rapidly eroded. But as the new technologies of communication inevitably spread deeper and deeper into

> the new societies it becomes ever harder to maintain local cultural autonomy. The paradoxes of dependence multiply and a political backlash results, of the kind which we are witnessing today in the international debate over the flow of news.[38]

As we shall see later in Chapter 9, the situation is highly complex and, even with good will on the part of all leaders, may be very difficult to change. It constitutes a positive challenge to all people of good will who dedicate themselves to promoting genuine peace rather than to Western strategies of pacification.

Having considered the many problems involved in obtaining the news with accuracy and clarity, we turn to other forms of information that are important to modern life. The rapid development of technology in the field of communication may lead one to believe that information is becoming available with increasing ease and abundance. As we shall see in the following chapter the availability of information is limited by many factors.

QUESTIONS FOR DISCUSSION

1. How did Teddy Roosevelt's insight and initiative benefit journalism?

2. Media editors find it difficult to present accurate newscasts each morning and evening. Identify several of their problems.

3. Monitor your newspaper for a week to discover examples of cases which seem to be news distortion.

4. Identify cases where disinformation formed attitudes and opinions of Americans.

5. When news is suppressed or distorted, how does the public interest suffer?

6. The section, "Rigging and Slanting the News," listed five criteria for choosing news stories. If you were editor which criteria would you use? Why?

7. How do the ordinary American newcasts of events occurring in developing countries affect Americans? How do they affect citizens of those countries?

Chapter 6

AVAILABILITY OF INFORMATION

Harry Truman once remarked that he looked with commiseration over the great body of his fellow citizens. For, reading newspapers, they live and die believing that they have known something of what has been passing in the world in their time. Truman's attitude may sound cynical but it seems nearer the truth than we may want to believe.

Previous chapters have treated the problems involved in selecting the news and presenting it in an accurate, clear, and balanced way. A more basic problem lies in the very availability of news and vital information. Put in different terms, our problem in this chapter is: while information is increasing at an unprecedented rate, why is its availability becoming more and more restricted to the ordinary person?

Explosion of Information

Jefferson and other founding fathers were convinced of the importance of information if U.S. citizens were going to act democratically. So in the Bill of Rights they insisted on the freedom of the press. Paradoxically, although the discovery of telegraphy in 1844 allowed information to flow more widely, it also empowered certain people to restrict that flow so indispensable for democratic action.

Media analyst, Neil Postman dates the beginning of a cultural shift in communication with the invention of the telegraph in 1837. Telegraphy created a new situation which later technologies simply intensified. Broadcasting information bits detached from their events, the telegraph cut the link between information and action. It made the relationship between information and action both abstract and remote. For the first time in human history, people were faced with the problem of information glut.[1] Moreover, the information was transmitted out of context and therefore people could no longer take appropriate social and political action.

The telegraph changed the traditional way of communicating within an action-oriented context. By depending on telegraphed news, people gradually became accustomed to cut information off from its bearings on decision-making and socio-political action. Thereafter, novelty and arousal of interest--not key issues or moral leadership--were what made information or news valuable. In this way, the telegraph transformed bits of information into commodities, into "things" that could be bought and sold irrespective of their practical democratic meaning or their utility for solving socio-political problems[2]

Four years later, the Associated Press was founded, and news from apparently nowhere, addressed to no one in particular, began to crisscross the nation. Wars, crimes, crashes, fires, floods--much of it with little social or political value--became the content of what people called "the news of the day." Soon the fortunes of newspapers came to depend not on the quality or utility of the news they provided, but on the quantity, the speed, and the distance related to the news[3]

During the first half of the twentieth century, two major wars sparked a vast amount of research in science and new technology. The sciences of biology, chemistry, physics, and astronomy became sub-divided into special areas and interlocked with one another creating new information at unbelievable rates.

This vast expansion of knowledge affects every moment of our lives in ways we rarely have time to consider. Our culture, life-style, life-expectancy, and our physical and mental well-being have all changed drastically--most frequently for the better. Unlike material possessions, the sharing of knowledge does not diminish what the donor possesses. On the contrary, the more that knowledge is shared the richer all humanity becomes.

Today, unfortunately, access to these riches often depends on computer literacy. The card files in some libraries are becoming obsolete. References, cross-references, abstracts, shelf lists, and periodical indices yield their information to those familiar with computers. Soon, those without computer literacy will be almost as handicapped as non-readers in a print-culture. Already caution flags arise from government studies. In 1996, in public schools nationwide, only 39% of black students, compared to 56% of white students, had access to computers at school. At home, only 15% of black families had a computer, unlike 36% of white families.[4]

Educational institutions are especially challenged to provide the skills needed to tap into this store of information. These skills must be made available to rich and poor alike or the existing gap between these two groups will continue to widen.

The acquisition of computer skills does not guarantee that we can obtain all the information we might need at a particular time. Unfortunately, our government and corporations keep much information from flowing freely to American citizens. In spite of the Freedom of Information Act, which many Americans feel protects their right to information, a secrecy mania has been developing at an alarming rate in government circles. Thoughtful, concerned citizens may well ask, "Just how available is information from the government?"

Government Restraints on Information

The classic case of the U.S. Army's massacre of 128 Vietnamese women and children at My Lai 4 may briefly have alerted some Americans to government restrictions on information. For 18 months the Pentagon refused full disclosure of that gory event.[5]

Several years earlier, however, and far closer to Americans than Vietnam, a process was harming their own bodies but knowingly left undisclosed. In 1966, General Electric discovered that their color TV sets were emitting X-rays at 100,000 times the recommended standard of 0.5 millirems per hour. 154,000 of these sets had already been purchased. Instead of alerting the public and recalling the defective sets, GE began trying to locate the sets and repair them. GE did report the defect, however, to the U.S. Public Health Service's National Center for Radiological Health. The NCRH also decided to remain silent but urged GE to make the announcement. The public remained ignorant of the danger posed by these sets until the information was leaked to a *New York Times* reporter. By this time unwary TV owners had been bombarded for months by X-rays that could cause mutations of their genes. Then Dr. Harold Stewart, the director of NCRH X-ray exposure lab, tried to downplay the dangers.[6]

In order to avoid similar incidents, the Freedom of Information Act (FoIA) was passed in July 1966. It requires government agencies to publish in the Federal Register "descriptions of its central and field organizations, . . . and the established places at which . . . the public may obtain information or make submittals or requests."[7] Each agency is further directed to:

> . . . publish or make available to public inspection all final opinions or orders in the adjudication of cases (except those required for good cause to be held confidentialand not cited as precedents) and all rules.. . . [Moreover] matters of official record shall be made available, in accordance with published rule, to persons properly and directly concerned, except information held confidential for good cause found.[8]

The exception clauses opened the way for numerous restraints on information and these exceptions to the Freedom of Information Act have been multiplying. Recently, journalists praised Attorney General Janet Reno for initiating a complete review of all FoIA regulations, exemptions, and policies in an attempt to make information more available. This review especially concerns the millions of documents that never should have been classified[9]

By the mid-1990's, the Freedom of Information Act annually brought more than 600,00 requests for material to federal departments and agencies. Delays of over two years in receiving the requested material often makes the material of no use to the requester. Senator Patrick Leahy comments: "Such routine failure to comply with the statutory time limits (10 days) is bad for morale in the agencies and breeds contempt by citizens who expect government officials to abide by, not routinely break, the law."[10]

Government restraints on information also affect the usefulness of libraries. Americans expect libraries to be a source of information, limited only by the size of the library and its budget. Yet libraries do not always have access to government documents that should be available to the public. In 1979, during the midwinter meeting of Governments Round Table, (a committee of the American Library Association), a new member mentioned the difficulty she had obtaining federal documents for her library. These documents had been readily available in the past. Two years later, the American Library Association (ALA) began keeping a log, "Less Access to Less Information by and about U.S. Government," listing government restraints on information.[11] For the three-year period, 1981-1984, the log listed 71 actions that the ALA believes threatened the free-flow of information to the public. "Less Access to Less Information #2," for 1985 and 1986, highlighted the trend to even tighter restraints. The material covers 34 pages and lists 110 actions.[12]

In its *Chronology* for January-June 1991, ALA maintains that:

> . . . a combination of specific policy decisions, the administration's interpretations and implementations of the 1980 Paperwork Reduction Act (PL 96-511, as amended by PL 99-500) and agency budget cuts have

significantly limited access to public documents and statistics.[13]

ALA has documented that since 1982, one fourth of the government's 16,000 publications have been eliminated.

Many librarians recognize that their increased difficulties in obtaining information started with the beginning of Reagan's first presidential term. At that time, the Office of Management and Budget was elevated to a superagency which was to controll the plans, programs, and information activities of all other agencies. Circular A-130, entitled "Management of Federal Information Resources," requires maximum reliance on the private sector to disseminate government information, the cost to be covered by a users' fee.[14]

Sooner or later many Americans may be priced out of the market--barred because poor from the benefits promised by the new technologies because they cannot afford to pay for them. Then the affluent would become even better informed than they are today while the less wealthy would become still less well informed.[15]

The less affluent often depend on public libraries in their search for information. The growing tendency in federal agencies to use computer and telecommunications technologies for the collection, storage, retrieval, and dissemination of data often limits its use. Contracts are arranged with commercial firms at the taxpayers' expense and often the information collected is in electronic format only. In the long run, automation will result in savings. Will information be available, however, to those who cannot afford a computer or pay for computer time? The ALA sees a ray of hope: "Now that electronic products and services have begun to be distributed to federal depository libraries, public access to government information will be increased."[16]

In the meantime, the executive branch of government itself annually makes 6.8 million documents secret and the number is increasing each year. Richard O'Curry remarks in *Freedom at Risk: Secrecy, Censorship and Repression* in the 1980's:

> The policies of the Reagan Amininstration set radical departures from the past. This is revealed not only in the comprehensive scope of the administration policies, but in its ability to institutionalize secrecy, censorship and repression in ways that will be difficult if not impossible to eradicate.[17]

Examples of tactics used to block citizens' access to information include:

1) In October, 1981, the Administration proposed exemptions from the Freedom of Information Act for trade secrets, financiali nformation, desk calendars of government officials, phone logs, and the minutes of meetings. The CIA was given complete exemption and told to close all case files for seven years.
2) The Administration eliminated programs that collected data from industry crucial to the health and safety of the public--information the public has a right to know.
3) Government publications that did not follow the Administration's pro-industry line have also been eliminated. Publications like *Car Book*, listing the results of government auto safety and mileage tests, are no longer published. These publications are cost effective as they provide consumers with the knowledge to settle disputes that otherwise might have to be settled by government agencies.[18]

Besides the exemptions that make availability of information difficult, the vast number of requests to government agencies from private individuals, especially prisoners, scholars, historians, and free-lance writers, as well as the media and other organizations, has created a huge backlog. As of May 31, 1996, the FBI had a backlog of 15,259 requests with an estimated 5.4 million pages of documents to be reviewed. About 248 of these requests were project cases covering 3,000 or more pages.[19] The 1996 Electronic Freedom of Information Amendment, sponsored by Senator Patrick Leahy and Senator Hank Brown, has passed the Senate. This bill will provide that information in electronic records be supplied electronically, and will speed up the time of response to requests.[20]

In the meantime, privatization of government information is having another effect, according to James Ridgeway, writing in the *Village Voice*: "The result has been to slowly cripple the functions of government that we take for granted." He cites as examples the probability that the best source of information on the United States economy is Japan. Besides their statistics on their own economy, the Japanese make their own informed estimates on how the United States operates. Also U.S. business now relies on Canada for data on cross border trade with Canada. When the Joint Economic Committee of Congress wanted to make a study of interest rates, the data was unavailable from the Treasury computer and the committee had to buy the information from a private company.[21]

We may hesitate in accepting I. F. Stone's unqualified statement that every government is run by liars, and nothing they say should be believed.[22] Too many instances, however, can be cited where information to the public was not only withheld but deliberately falsified.

Writing in the *Columbia Journalism Review*, Anthony Marro relates how FBI director Clarence Kelley was asked in 1975 about FBI break-ins without warrants. He replied that some warrant less break-ins by FBI agents had taken place in the past but only in cases against foreign espionage and counterintelligence agents. He assured his questioner that J. Edgar Hoover had ended this kind of activity in 1966. As a matter of fact, however, there had been thousands of illegal break-ins, before and after 1966. Most of them were against American citizens, of whom many had never been charged with any crime.[23]

Anthony Marro comments further:

> While it is not clear that the Reagan administration is any more duplicitous than others, it unquestionably has gone well beyond other recent administrations in its attempts to bottle up information, to prevent public access to government officials and records, to threaten and intimidate the bureaucracy in order to dry up sources of information, and to prevent the press and public from learning how their government is functioning.[24]

As an example, Marro cites the reply of White House spokesman Larry Speakes at a press conference on October 24, 1983. A reporter had asked whether U.S. troops had landed in Grenada. Speakes hesitated, checked with a member of Reagan's National Security Council, and answered, "Preposterous!" He went on to deny that any invasion was even being planned. The landing took place the next day.[25]

Corporate Restraints on Information

While living with the U.S. government restrictions on the flow of information, libraries also experience how media corporations are exerting increasing control. These global giants are not satisfied expanding their control over the technological systems that gather and disseminate news and information. They and their subsidiaries are also amassing world copyrights on decades-old information and popular culture. In this way, they gain control of news archives, magazines, books, TV programs, film libraries, and musical compositions. By owning the copyrights they can demand a fee for the use of any of these materials. Even important statistics now in data banks are available only for a fee. Meanwhile, subscription costs for scholarly journals have greatly increased.[26] If this trend continues, less and less information will be available except to the privileged few who have sufficient funds to pay for what they need.

Technological know-how and knowledge of markets and resources are increasingly becoming the basic product of modern industry. Boundaries between markets are breaking down and the world is evolving toward one integrated commercial system. Therefore, the processing of information--the ability to acquire accurate data and transfer them quickly from one corner of the globe to another--is becoming ever more critical to business success. Information becomes the single most important element in any program to control or eliminate private cartels. On the global level, nations should cooperate to gather and release data on business practices that affect trade between them. Among practices adversely affecting such trade, cartel schemes rank first.[27]

Perhaps the most important information that needs to be made public is data on ownership. One of the most astounding factors in the U. S. today is that no one, not even the government, can obtain the information necessary to answer the most basic of economic questions: Who owns and controls the nation's industrial and financial assets?[28]

One might think privately-owned businesses too small to bother about. Although this attitude is generally on target, one rightly wonders whether companies beyond a certain size should be required to give a detailed public financial account. For example, with annual sales of $47 billions in cereals, salt, petroleum, etc., the privately-owned giant Cargill makes a mighty impact on the national and global economies. Should it continue to be largely exempt from public disclosure of its inner workings?

In 1979, in an attempt to gain some hold on the financial situation, the Security & Exchange Commission (SEC) did tighten its requirements about disclosure. A publicly held corporation must now reveal if any of its officers, directors, or shareholders own 5% or more of any class of stock. Five percent is often enough to control the firm. This regulation, however, has so many loopholes, that no one can learn who really controls the firms. One well-known way to conceal a large interest is to disperse the stock among members of the family so that no one person appears to surpass the 5% but together they control the firm. Other techniques for control include front organizations and legal representatives. Often four or five layers of such camouflage are employed, thus shielding the true owner.

Along with data on ownership the public needs information about the internal workings of transnational corporations. Some ways multi-national corporations use to withhold information are, when discovered, rightly resented by the public. For example:

> Oil companies do not want the public to know the low rate of taxation on their revenues. They do not want to disclose that they are spending their profits on such acquisitions as department stores instead of finding ways to make the United States less dependent on foreign oil.[29]

Several kinds of internal corporate data should be made public--among them financial records and tax returns, now regarded as privileged information. These corporations do not want it known that several sets of books are kept, each presenting a different picture of reality and each designed for a different audience.[30]

Military Restraints on Information

At the end of World War II, the American people were aware that a new and terrible weapon had been used but they did not receive information about the tremendous human suffering it had caused. Public opinion was manipulated by the control of information by government officials and by the elation of victory. Americans viewed the atomic bomb as a means to bring their husbands, sons, and brothers home. They never questioned the "illegality of the Manhattan Project or the lack of Congressional approval of the spending on the bomb or the decision to use it."[31]

Safety and health risks from nuclear and military accidents are another area where the public is poorly informed. Since the end of the Cold War, the false alerts of a potential nuclear attack may not seem as important as they once were. Nonetheless, the question remains, how many people realize that in the 18-month period from January 1979 to June 1980, there were 3,703 false alarms, 152 of which were serious enough to have represented a potential attack?[32] More relevant for us now, how many false alerts occurred during the past year and how widely publicized was this vital information?

Accidents with nuclear weapons remained classified information for years. According to an unclassified document of the Department of Defense, there were 32 accidents involving United States nuclear weapons between 1950 and 1980. Many of these accidents involved equivalent to 24 million tons of TNT. For example, on January 24, 1961, structural failure in a B-52 carrying two nuclear weapons over Goldsboro, North Carolina, caused the break-up of the plane. The bombs were both twenty-four-megaton bombs, which is equivalent to 24 million tons of TNT. One bomb crashed and five of its six safety catches were triggered. Only one safety catch prevented the destruction of much of North Carolina.[33]

Dr. Helen Caldicott, a physician who has studied in depth the effects of nuclear radiation, writes in *Missile Envy*, "The Stockholm International Peace Research Institute estimates there were 125 U.S. nuclear-weapons accidents, major and minor, between 1945 and 1976, or about one every two and a half months."[34]

Considering the proliferation of nuclear weapons since 1976, a conservative estimate to date would be at least 210 nuclear-weapons accidents. In the mid-1980s, the Navy admitted the existence of classified documents, dated from March 1973 to March 1978, entitled "Summary of Navy Nuclear Weapons Accidents and Incidents." These documents are hundreds of pages long.[35] How much of this information has been made available to the public? Surely, the public has the right to know when a large segment of the population is, or was, in danger of being wiped out. With such vital information withheld, how can the democratic public be expected to make political decisions that are informed and intelligent?

University-Industry Partnerships

Beyond the yoking of university research to various government -sponsored projects, an additional practice is growing between universities and industry which also restricts the availability of information. Traditionally, universities tend to research scientific issues and test scientific theories. The results of university research are often published in scholarly journals or presented at scientific conferences. In this way, such information is made available to the scientific world for further research and refinement.

On the other hand, industries usually conduct research to improve old products or test new ones. The goal of this research is to market a product that will enhance the profits of the industry.

The National Academy of Sciences' report, entitled "Industrial Perspectives on Innovation and Interactions with Universities," hints at what university-industry alliances involve. These kinds of partnerships are becoming more and more common--for instance, the one announced in June 1991 between Proctor and Gamble (P&G) and the University of Cincinnati (UC).[36] Although details of the partnership have not been released, convergent trends of the past decade point to the area of research. P&G has been buying up drug-producing companies and UC has made no secret of the fact that in terms of research money it wants to be one of the country's top 20 medical schools.

Frank Tepe, UC's vice president for research, explains the usual arrangement for partnerships of this kind. He says, "In most cases, the company funds basic research by the university, which can license the rights to any potentially marketable discovery. The company has first rights to the license and markets the product."[37] Since the mid-1980s, UC has developed a technology transfer office to see that its researchers' discoveries are transferred to the marketplace only through some corporation's first option. Once patented, this information, created by university research, is thus kept from public dissemination and widespread usage.

Casey Kiernan, the senior program officer for the National Academy of Sciences, is well aware of the dangers in the present trend of industry-university partnerships. He said, "In some instances, universities do research that they might not have done otherwise--more commercially oriented. That worries some people, because that changes the culture."[38]

Making ownership of such information private reduces the availability of the information to the public or to the government. Even though this commercial use of information has become a central force in our society, there has been very little discussion of the political and policy issues involved. There is a real struggle between those who wish to appropriate the country's information for private gain and those who favor the fullest availability. The trend toward more and more private ownership of information is a danger to a free society.

Availability of Private Information

By contrast, the only information that is becoming more available is information about individuals. This information is compiled by combining government information from census forms, license applications of various kinds, and social security data, with data from insurance and credit companies.[39] As they approach their 65th birthday, senior citizens are sometimes shocked to be deluged with birthday greetings from insurance companies offering supplemental insurance. At any age, it is disconcerting to move and discover how soon all the junk mail follows you to your new address. You may well feel that somehow your privacy has been invaded in spite of laws on the books to protect it.

For legal purposes, privacy is defined as a condition in which individuals, groups or institutions can determine for themselves, when, how, and to what extent private information about them is communicated to others.[40]

Apparently, privacy laws have as many exemptions as the Freedom of Information Law. For example:

1) The Fair Credit Reporting Act passed in 1970, bars credit agencies from sharing information with anyone but authorized customers. Although the word authorized seems restrictive, various categories of customers create exceptions.

2) The Privacy Act of 1974 bars Federal agencies from sharing information for a different purpose from which it was gathered. This law has many exceptions.

3) The Right of Financial Privacy Act of 1978 sets strict procedures for Federal agencies who are tempted to rummage through a customer's records in banks. The weakness to this law is obvious when one realizes that it does not apply to state and local governments.[41]

It is ironic that an individual's video rental records are better protected than one's medical or insurance files. Passed in 1988, a Video Privacy Protecting Act prevents retailers from disclosing video-rental records without the customer's consent or a court order. Advocates for greater protection of individual's privacy are pushing for similar laws for medical and insurance files. Most citizens have little idea that files on them exist and individuals can hardly learn how these records are being used.[42]

Karen Hockman was blissfully ignorant of the extent of the files on her until she received a call from ITT encouraging her to subscribe to their long-distance service. Hoping to rid herself of the unwanted caller, Karen said she did not make many long-distant calls. She was startled and angered by the reply, "I'm surprised to hear you say that. I see from your records that you frequently call Newark, Delaware, and Stamford."[43]

Telephone companies are not the only invaders of privacy on phone use. Police recently discovered two persons operating a cellular phone number theft. They found 43 cellular phones, 6 computers, and the printout of 80,000 stolen numbers of phones in New York City. The numbers were for sale for $500 each. It is uncertain how many of the numbers were sold but New York Police Commissioner Howard Safire was one of the victims of the scam. He received a bill for $5,000 for calls to the Middle East which he had not make. His bill was just a small part of the $1.5 million phone fraud amassed each day nationwide. Cellular phone companies are taking steps to prevent fraud by a process called "authenticating." Encrypts are added to the phone number making it almost impossible to replicate programming of a cellular phone.[44]

A French hacker, however, did decrypt a message that had been posted on the Internet as a challenge. He worked eight days with 120 workstations and a few parallel supercomputers. He used approximately $10,000 worth of computing power to break this single message. The cost of breaking the encryption ensures the future security of phone numbers and of credit card numbers with their expiration dates and their limit of credit.[45]

Small marketing companies that can afford the fees gain a competitive edge from data obtained from information brokers. Three major credit bureaus--TRW in Orange, California, Equifax in Atlanta, and Trans-Union Credit Information in Chicago--have 400 million records on 160 million individuals. From TRW's Financial Lifestyle Data Base, a business can buy lists of names, addresses, and phone numbers of individuals who have been categorized by income, credit cards, and the amount of credit available. From Equifax, one can purchase a computerized breakdown of certain towns and neighborhoods with profiles of residents' spending habits.[46] After collating and studying all this information, a company can pinpoint prospective customers and do a massive direct mailing promotion of their products. A follow-up phone call to choice prospects will often result in new customers, increased profits, and the opportunity to expand the business.[47]

Car dealers rely on credit information supplied by information brokers in negotiating sales for new cars. Astute customers never give their last name until they feel they have been quoted the best possible price by a dealer and are ready to close the deal. They know how car salesmen leave their customers for a few minutes, quickly check their financial status on the computer and deal accordingly.

Not every small business can afford the fees of the TRW, Equifax, or Trans-Union Credit Information. Instead they go to information brokers who have catered to small businesses since the 1980s. For instance, Mark Hanna negotiates with local credit bureaus to buy reports for $5.00 each--about three times the price they charge major banks and retailers. Mark then resells the reports for $10.00 each to such people as lawyers and small business owners who need only a few reports and find this way cheaper than joining the credit bureau. Mark now has thousands of customers in all 50 states.[48]

Information collected by data brokers can be used by criminals, as Jacob Maarse of Sierra Madre, California, discovered. Jacob owned debt-free land and was one of the targets of a scam by Robert Fine, Jr. While serving time in prison for mail fraud, Fine obtained the names of owners of debt-free land, such as Jacob. He contacted information brokers and got these owners' credit histories, former addresses and the numbers of their Social Security Cards, credit cards, and drivers' licenses. With this information he applied for loans amounting to $200,000 by posing as Jacob Maarse and using Jacob's land as collateral. Fine told the lenders he needed the money for AIDS research.

Fortunately for Jacob, two of his friends worked at companies processing the loans. In a casual conversation, they discovered that they had both been approached for a loan by a person claiming a near break-through in AIDS research. Further investigation revealed that Fine had collected one million dollars from other land owners before he was discovered. Stephen Mansfield, assistant U.S. attorney in Los Angeles, expressed the reaction of many people when he said, "What I found shocking about this case was the ease with which anyone can obtain anyone's credit history for $10.00 or $15.00." Perhaps even more shocking is the comment of Jean Noonan, top credit regulator for the Federal Trade Commission, "We are spending millions of dollars trying to convict people on narcotics charges. You can understand that this [credit scam] would be a relatively low priority for most federal prosecutors."[49] One wonders, would victims like Jacob Maarse really understand?

The interconnecting of computers into networks can lead to the production of "psychographic profiles." Although databases may be harmless when separated, when linked together they can be damaging, perhaps even incriminating. Information about an individual can be transferred from one organization to another without the individual's authorization or even knowledge. Increased storage capacity of more powerful computers encourages the tendency to retain outdated and no longer accurate data on individuals. When this data is included in the psychological profile it may be very incriminating. The following partial profile gives some insight into the power of these networks, the amount of detail included, and the innuendos drawn from the data.

NATIONAL DATA BANK

DAILY SURVEILLANCE SHEET

Confidential

July 11, 1987

Dennie Van Tassell
San Jose State College
Male , Aged 38
Married
Programmer

Purchases	Wall Street Journal	0.50
	Breakfast	2.65
	Gasoline	10.00
	Phone (328-1826)	0.10
	Phone (308-7928)	0.10
	Phone (421-1931)	0 .10
	Bank (Cash Withdrawn)	(120.00)
	Lunch	2.00
	Cocktail	1.00
	Lingerie	21.85
	Phone (369-2436)	0 .35
	Bourbon	8.27
	Newspaper	8.10

COMPUTER ANALYSIS

Owns stock (90% probability)

Heavy starch breakfast. Probably overweight.

Bought $10.00 worth of gasoline. Owns VW. So far this week he has bought $40.00 worth of gas. Obviously doing something besides just driving the 9 miles to work.

Bought gasoline at 7:57. Safe to assume he was late to work.

Phone No. 328-1826. Belongs to Shady Lane. Shady was arrested for bookmaking in 1972.

Phone No. 308-7928. Expensive men's barber--specializes in bald men or hair styling.

Phone No. 421-1931. Reservations for Las Vegas (without wife.) Third trip this year to Las Vegas (without wife). Will scan file to see if anyone else has gone to Las Vegas at the same time and compare to his phone call numbers.

Withdrew $120.00 in cash. Very unusual since all legal purchases can be made using the National Social Security credit card. Cash usually only used for illegal purchases. (It was previously recommended that all cash be outlawed as soon as it is politically possible).[50]

Concerning this type of profile, the Attorney General of New York declared:

> These profiles and projections based on them, could be used by credit companies, government officials, landlords, insurance companies, investigators, marketing firms and others for decision making. Unauthorized disclosures of psychographic profiles and projections could seriously harm personal and professional lives and the emotional and psychological well-being of the subjects of disclosures.[51]

In spite of Jean Noonan's comment quoted above, can the Federal government continue to ignore the proliferation of such potentially criminal activity?

The availability of information is not only a national problem but has international implications. Foreign governments are increasingly concerned about the vast amounts of computer-based digital bits moving in and out of their territories. This data affects their economies and frequently their political interests. To counteract this situation, Europe and Canada are setting guidelines to regulate transborder data flow. The Canadians especially are concerned about the vast amount of information on their citizen's private lives stored in data banks in the U.S. In Europe, most countries have already developed strict restrictions on the electronic transfer of personal information about their citizens to other countries.[52]

In this chapter, we have seen how government and corporations restrict information and yet how loop holes are allowed for increasing private information about individuals. As we shall see in the next chapter, the role of government and big business extends beyond influencing news and information to a broader control over the various media themselves. Furthermore, the hierarchies in these media determine the attitudes and values to be presented to their audience. In Chapter 11, we will give some suggestions how private individuals and groups can exercise their freedom of choice toward the media.

QUESTIONS FOR DISCUSSION

1. How did the telegraph and later technologies change the kind of information public received?

2. Suggest practical steps to make government documents more available to the public?

3. Does restricting the availability of government information at public libraries affect you? How?

4. By amassing old copyrights and tightening control on new ones, how do media companies affect our lives and our culture?

5. Do ordinary citizens have a right to know about accidents with nuclear weapons? To what changes would publishing such information lead?

6. Does a university's mission of freely researching for new knowledge get sidetracked when yoked to industry or government projects?

7. Has your privacy been invaded electronically? In what ways?

8. Is the government responsible to help protect your privacy? How can it do so without getting too much control over your life?

Chapter 7

ROLE OF BUSINESS, GOVERNMENT, AND THE MEDIA HIERARCHY

In 1927, John Dewey, an insightful philosopher, realized that the indispensable role of the media was to distribute knowledge freely and fairly. For democracy could not survive unless the media nourished public opinion and stimulated the subsequent discussion of issues. Dewey also believed that the American people had often been inadequately informed and misled by the media. So he realized that if monied interests came to control the media and if the people failed to resist, then a free democratic way of life would decline into that of a slave.[1] Long before the rise of TV and electronic news services, Dewey sensed how greed tampers with public opinion and how it disregards the common good.

The world of big business and government works, of course, according to the wrestlers' rule of "squeeze before getting squeezed." This world feeds on fast deals, sweet perks, and special promises, like the world of professional wrestling itself. Money has muscle here and sometimes leads these giant wrestlers to brush little folk aside and to crush meddlesome resisters underfoot. In this tough arena, then, business and government people cannot be too squeamish about having dirty hands. According to Carl Friedrick, a Harvard political scientist, the politics of government must work with the "diseased functions" (or pathologies) of propaganda, secrecy, corruption, betrayal, and violence.[2] Big business must do the same. The only reasonable way to keep these pathologies within limits is, as Friedrick also saw, to employ countervailing forces.

Constantly surrounded by greed-driven business firms and governments, mass media find that their life, too, lies in "squeezing before getting squeezed." For the most part, however, media people are neither conspirators working behind the scenes with hidden agenda and power plays nor heroic defenders of the rights of the downtrodden.[3] Rather, media people and their institutions are, like most of us, a complex interweave of power and limits, of virtues and vices. So, it helps to recognize the media's achievements as agents of liberation before highlighting the real domination of, in, and by the media. So, before entering the main theme of this and the next chapter--the real domination of, in and by the media--two preparatory steps are needed: 1) to recognize the media's achievements as liberating yet endangered agents, and 2) to gain a clear picture of the media world with their giant players and trends.

The Media as Liberating yet Endangered Agents

The media's news and weather reports, documentaries, and even advertisements supply the uninformed with information needed for making sounder and freer choices. For instance, the media have supplied early warning of hurricanes, pointed to new reliable medicines, and highlighted the danger of illegal drugs. They have also made the threat of a population explosion become more widely recognized round the globe. As a result, the birthrate in many countries has shown a downturn, even if the battle is not yet won.

Moreover, the media regularly present the various sides of a question so their audiences may more readily find a fuller picture than they would have gained if offered a single-sided presentation. For example, by informing people about the strengths and weaknesses of political candidates, the media provide potential voters with a more informed freedom of choice. The media also promote hands-on democracy by fostering call-in programs, sampling opinions in the street, and creating shows that involve interactive participation.

Many freedoms of Americans--their freedoms of democratic discussion, assembly, election, and government, along with their economic and religious freedoms, largely depend upon First Amendment rights protected by freedom of the press. So, when the media continue to battle to protect freedom of the press in and out of court--often in inconspicuous ways--they are achieving something far more significant than simple self-defense. Such are a few ways in which the media promote the freedoms of people and deserve their thanks.

Aware of these powerfully liberating services, some media directors may also realize that because their institutions exercise such a vast influence, their very height and weight offer especially vulnerable targets. Symbolically, they fancy to themselves that their own seemingly solid metropolitan towers of power rest as much on shifting sand as upon solid rock. They fear that if someone discovers previously unearthed facts or if scientists make a technological break-through, they will be forced to retract their assertions or revise their editorial policies, or be swallowed up by a still bigger merger. They feel uneasy when classified as a "part of the Establishment" by environmental and religious groups or even by a majority of Americans who protest certain positions they have taken. Still more disconcerting, they cannot yet discern which changes the incoming wave of the electronic revolution will demand of them.

High level media directors also know that, at least until the present, they have escaped periodic legal scrutiny. For as yet, no law forces them to account regularly before an independent forum of national and international referees about the policies and practices they have employed toward their audiences and subordinates. Media managers also feel nervous before the recently emerging and intensely competitive economic blocks, like the European Union, Japan, and OPEC, which bump and bruise the whole U.S. economy and themselves with it.

What media directors may most fear, however, is the American people. For directors cannot forget how quickly this people's views on Vietnam turned around. They realize that Americans have enough common sense to put up with a limited amount of half-truths and fluff and will believe only so much. Moreover, Americans can raise certain questions embarrassing to the media, for instance, their silence about monopolies or their promotion of environmental or international awareness. More than a few Americans also have feelings easily enough aroused that they may simply disconnect themselves from TV or other kinds of media.

All these factors make media leaders feel how shaky is their position and how limited their power. Yet they recognize the grounds for their conviction that their daily services inject a profoundly valuable asset into public life.

The next preparatory step is to gain a clear picture of the media world itself, one that identifies its major competing giants.[4] Using 1995 as a benchmark, one asks fittingly: Who are the media giants in this wrestling ring--the big broadcasters, cable operators, and those that do both?

Which broadcasting companies hold cross-ownerships in the business of publishing newspapers, magazines, records, and books? What are the various teams that support these media producers--that is, the largest program producers, furnishers of technology, agencies, and providers of satellite services? If this picture is brought into clear focus, the reader will grasp the complexities of this world of intra-competing media.

Giants in the Media World

Among broadcasters, the "big ten" are Capital Cities/ABC, CBS, General Electric, News Corporation, Tribune, Westinghouse, BHC Communications, Gannett Co., Scripps Howard, and New World Communications.[5] Their revenues in 1993 ranged from $ 4.6 billions for Capital Cities/ABC down to $369 millions for New World. Interestingly, in 1993 "commercial broadcasting had total advertising revenues of approximately $ 33.3 billion," with TV capturing 71.1% of that income.

Cable has mushroomed into a competitive nemesis for broadcasters, capturing ever larger percentages of the info-entertainment market. Currently there are more than 11, 800 cable systems operating in the U.S. In 1995 the ten giant cable firms were: TCI, Time Warner, Continental, Comcast, Cablevision Systems, Knight-Ridder, Multimedia, Times Mirror, Cablevision Industries, and the Washington Post. A hint of cable's rapid growth is given in their revenues for 1993, which ranged from $ 4.2 billions for TCI down to $ 363 millions for the Washington Post. In 1994, the cable industry's overall revenue totaled $ 23 billion.

Which companies are the super-giants that own both broadcasting and cable systems? Leading the pack of the fifty largest multiple systems operators--MSOs as they are called-- are the following ten firms listed with the millions of their total basic subscribers:

Rank	MSO	Total Basic Subscribers
1	Tele-Communications Inc.	10.2 (millions)
2	Time Warner Cable	7.3
3	Cox Communications Inc.	3.0
4	Continental Cablevision Inc.	3.0
5	Comcast Cable Communications	2.8
6	Cablevision Systems Corporation	2.6
7	Rogers Cablesystems Limited	1.9

8	Adelphia Communications	1.4
9	Cablevision Industries, Inc.	1.4
10	Jones Intercable, Inc.	1.3

If we expand our view to take in the print media, what was the situation of cross-ownership in 1995? That is, how many of these broadcasting firms or cable systems or multiple system operators were also publishing newspapers, magazines, books, and recordings in 1995? By then these cross-ownerships had sprouted up almost everywhere. For instance, Capital Cities/ABC owned 14 major daily newspapers plus 69 community newspapers, 17 broadcasting stations, along with financial services and a medical group. As for Gannett, besides its flagship, *USA Today,* it then owned 33 major dailies and 88 community papers, as well as 17 broadcasting stations. The Hearst Corporation then owned 11 daily newspapers, 15 business publications, and 9 radio-and-TV broadcasting stations, besides being heavily invested in overseas publications. Multimedia owned 15 newspapers, 6 broadcasting stations, and more than 125 cable TV systems. In the U.S., Scripps Howard owned 14 daily and 7 non-daily newspapers along with 9 TV stations, while its international division owned 11 more newspapers abroad.

Undergirding all these cross-ownerships and multiple systems operators, what are the support systems that keep the media world alive? Broadcasters and cable directors cannot operate without programs, so our focus shifts to the ten major producers of programs. In 1995 these were: Viacom, Turner Broadcasting, Liberty Media, Paramount, QVC Networks, Disney, King World Productions, Gaylord, Spelling, and International Family Entertainment. Their revenues in 1993 had ranged from $ 2 billions for Viacom down to $ 208 millions for International Family Entertainment.

Then, too, broadcasters, and programmers cannot function without the technology to carry out their work. So, a glance at the ten giants in technology is in order. In 1995 these were: Sony, Matsushita, General Instrument, Harris, Scientific-Atlanta, 3M, Tektronix, Kodak, Ampex, and Varian. Their revenues in 1993 ranged from $ 3.4 billions for Sony, to $108 millions for Varian.

Similarly, broadcasters, and programmers cannot function without the agencies that mediate actors' contracts, provide newswire services, supply weather reports, etc. *The Broadcasting & Cable Yearbook, 1995* listed the following 15 "Agencies and Services," of which only some are familiar household names: Interpublic, Saatchi & Saatchi, WPP,

Omnicom, Grey, Hughes, Dun & Bradstreet, Ceridian, Katz Communications, FCB, Reuters, United Video, Unitel, IDB, and the Associated Press. Whether well known or not, these supporting agencies and services gathered revenue in 1993 that ranged from $ 991 millions for Interpublic down to $ 58 millions for the Associated Press.

Finally, given the global nature of communications today, programs and news, stock reports and communications via phone, fax, and e-mail, depend heavily on owners and operators of satellites. Without considering military and other governmentally owned satellites, the major providers of other satellite services are AT&T, COMSAT Corporation and COMSAT World Systems, GE, and INTELSAT (International Telecommunications Satellite Organization). Without all of these players in the media world, we would have precious little sense of being a global village, of being almost as close to Beijing and Moscow as to our own civic center for receiving daily news and feeling their problems.

Absent from the foregoing picture are upsetting influences such as that exercised by Rupert Murdoch, the international media mogul. Rising out of debt in the early 90's, Murdoch bought Hong Kong-based STAR TV in 1993. This station has the potential of reaching the entire middle- and far-East; that is, three-fifths of the world's population. As owner of half of British Sky Broadcasting (BskyB), Murdoch recently upgraded it from 6 to 14 channels. He also pledged to buy Delphi Internet of Massachusetts, a gateway to the Internet and its 20 million computer users. Clearly Murdoch wants, as *Time* put it, "to cover the earth with his own digital superhighway."[6] Wielding such a global network, Murdoch possesses a global outlet for his own Fox Broadcasting Co. Through it he can propagate around the world the more than 2,000 movie titles which his 20th Century Foxfilms owns. Besides this, Murdoch's News Corp. empire has been built on tabloids that have thrived on sensationalism. He owns one-third of Britain's national daily newspapers--enough to raise the question whether democracy can flourish when one man speaks with so loud and omnipresent a voice. This objection is on target for the people of Britain, but may miss the people of China. For since China's repressive government cannot control electronic information and therefore not much of the radio and TV broadcasts into China, Murdoch's STAR TV may perhaps foster democracy in China. In brief, Rupert Murdoch is clearly a super-giant in the media world. Yet he is hardly the sole "wheeler and dealer" in the heady game of competition raging in the media world.

This nearly-current picture of the media world lets the reader sense the hurricane-like forces that are wrestling to squeeze and mold the digitally-based communication systems of the future. Recent legal deregulation of the media and the whirlwind of recent mergers, such as Disney-ABC, Gannett-Multimedia, Westinghouse-CBS, have concentrated still more power in still fewer hands.

Already before this merger madness of the last decade, even expert media researchers could hardly keep abreast of the spiraling phenomenon of mergers in the media industry. Yet the reader may get some feel of the present turmoil in the media world if he or she expands the above sketch to include business pressures on the media as of 1991. By contrasting the 1991 picture of the corporate world's impact on the media with the just-sketched 1995 picture of the media themselves, one can detect some of the major squeezes applied to the media by big business and the resultant trends in the media.

Having thus completed the two preparatory steps needed for context, one can more profitably ask which constraints big business and government apply to the media. For in this and the following chapter, our principal aim is to search for patterns of restraint upon, in and by the media. Guiding this search is a series of three main questions. First, how do big business and government control the media? Then, how does the media hierarchy dominate individuals and groups within the media system itself? And finally, in the next chapter, how within this context do the media invade the lives of their ordinary audiences? First, then, from the non-media world, what are the business forces pressuring the media?

Business Constraints on the Media

Back in 1991, ten business and financial corporations were tightly harnessing the media. Specifically, three quarters of the major stockholders of ABC, CBS, and NBC were banks such as Chase Manhattan, Morgan Guarantee Trust, Citibank, and Bank of America. Corporations like these owned the three major TV and radio networks (ABC, CBS, and NBC), 34 subsidiary television stations, 201 cable TV systems, and 62 radio stations. They controlled various motion picture companies like Columbia Pictures and Twentieth-Century Fox, 20 record companies, 59 magazines including *Time* and *Newsweek,* 58 newspapers including the *New York Times*, the *Washington Post*, the *Wall Street Journal*, and the *Los Angeles Times*, and 41 book publishers.

In addition, both then and today, media leaders often form interlocking directorates by selecting corporate executives from big business to be members of their boards of directors. Naturally, then, media leaders tend to share the biases of big business. They identify with the interests of business corporations more than with those of ordinary workers or of the unemployed. In fact, James Fallows, editor of the *Atlantic Monthly*, sees the result of the giant corporations' takeover of the media as now forcing publishers and editors to insure their investors of receiving no longer a modest profit but the "prevailing market return." He sees that the news business has been converted into "just another corporate operation, where whoever is in charge must be as driven by the demands of the financial market as their counterparts in the banking and steel-making and fast-food industries."[7]

In this light, one can better understand why in 1935 the Newspaper Publishers Association disgraced itself by condemning the proposed Child Labor Amendment. Because editors tended to favor big business, America's leading newspapers at first opposed food and drug regulation as well as legislation to assist consumer cooperatives. Publishers resisted both the Federal Trade Commission's regulations against fraudulent advertising and the rule of the Federal Communications Commission which restricted newspapers from owning broadcasting stations.[8]

Provocative media critic, Michael Parenti, seems to think that the situation John Dewey warned Americans about in 1927 is now being realized. For Parenti says, " . . . [in the American media] one confronts a precooked, controlled, centralized, national news industry that is in sharp contrast to the `pluralistic diversity' of opinion that is said to prevail in the U.S."[9] Crying out for a grass-roots movement against the control of the media exercised by a clique of corporate giants, Mark Miller describes this trust's downward pressure:

> The ubiquitous soft porn, the gangsta manners, the shock jocks and the now-obligatory shouting of the F-word are all products of the same commercial oligopoly that is also whiting out the news, exploiting women, celebrating gross consumption, glorifying guns and demonizing all the wretched of the earth.[10]

Then, too, established corporations--like the oil, auto, tobacco, and alcohol industries--put pressure on the media in order to tether most Americans to their products. These corporations threaten to terminate advertising accounts if the media broadcast too many investigative reports touching certain sensitive areas. For example, media do not show how feasible is an auto engine that gets 75 miles-per-gallon of gasoline. They do not campaign to have Caribbean and Central American sweatshops closed or Mexican workers' wages raised to perhaps half of the wages of their American counterparts. Nor do they readily divulge that American banks recoup more money in annual debt service to the U.S. from Third World nations than Americans invest in these nations through aid and development. If the media intend to stay in business, they may not reveal such lucrative trade secrets, regardless of the greater common good of the global human family.

Furthermore, big oil companies also wanted TV stations to keep showing the accident at Three Mile Island again and again. For if enough Americans began fearing a nuclear disaster, that would weaken support for research into safe nuclear and solar energy--research that just might break Americans' addiction to oil.

The current new order of corporate pressure upon the media became clear recently. A $ 10.2 billion auction of electromagnetic spectrum space revealed the giant size of the power-brokers at play. Of the five biggest corporations winning this auction, four were backed by Japanese and Korean investors.[11] Then, too, in 1995 the FCC simply dropped prosecution of Rupert Murdoch although he had fudged the actual foreign ownership of his firm. Again, on its *Day One* show, ABC News broadcast the truth that Philip Morris had manipulated the nicotine levels in its cigarettes. As a major advertiser through Kraft Foods, Philip Morris pressured ABC to apologize for broadcasting this factual truth. Moreover, when CBS was all ready to feature Jeffrey Wigand's whistle-blowing on Brown & Williamson for its addictive tobacco practices, CBS's in-house counsel ordered the old news hounds at *60 Minutes* to bury this explosive interview[12]. And, above all, as Mack C. Miller points out, "the media trust suppresses information and debate on *all* monopolies."[13]

Among corporations, PR firms and the ad industry play their large and distinctive roles in controlling the media. PR firms pressure the media to keep their own images bright and shiny, such as the integrity image of CBS or CNN or Time Warner--otherwise, ratings will drop. As for the ad industry, it has surely turned the media into a new breed of town criers, who are now forced to be more concerned about bandying abroad the advertised goodies of business than to present the news in an adequate context. For instance, the *New York Times* gives 60% of its pages to ads, 40% to news. Local newspapers, like the *Cincinnati Enquirer*, assign an even smaller percentage of their paper to news.

Advertisers have even co-opted educational Channel One so that it must show ads for toys and candy to captive audiences of children in primary grades.

Since children ages 2-11 spend or influence household expenditures of $100 billion every year in the U.S., and since these little consumers also watch as much as 26 hours of TV a week, the ad industry, especially via Fox, has seized the opportunity thus offered. Even if one bypasses the ads infiltrating educational Channel One to focus just on commercial TV's programs for children, one finds that the new competitors neither improve program quality nor decrease ads. One study summarized that "the regulatory philosophy of marketplace competition does not always provide the best solution for protecting young audiences as for-profit programming motives often conflict with children's best interests."[14]

Video news clips reveal another area of corporate pressure on the media: its power over a television station's busy newsroom. Most television stations currently operate under radical cost-cutting policies and their newsroom crews have more than enough jobs to carry out. So, by themselves, they find it hard to create by deadline time a high quality, well-packaged news clip with competent participants. Why not, instead, accept footage supplied by a corporation's PR department and simply air it as submitted? For these professionally produced "news clips" claim to cover both sides of a particular issue, even if they also carry a spin that serves corporate interests.[15] Does being a "free press" mean being free from such interests?

According to *Advertising Age*, corporations invested a staggering $ 150 billion on national advertising in 1994.[16] To create greater marketability for their sponsored products, the ad industry requires the media to reach the widest possible audience. So, the media have had to gear their programs to fit the average intelligence of a sixth grader--the rough measure used for mass audiences. As a result, the media often feel that they must address their audience in easily understandable terms--read, "unrefined and blunt." They must appeal to coarse levels of taste, not infrequently descending to that of *Beavis and Butt-head*. They must also keep capturing their audience's attention by appealing mainly to its animal interests. So, they flash on the screen verbal fights, scenes of violence or sexual allure, along with glittering baubles and whatever takes surprising turns. Finally, as mentioned earlier, the ad industry pressures the media not to criticize corporate interests too sharply lest income from ads be quickly shut off.[17] Overall, then, the media are in the grip of "an ever-tightening convergence of globe-trotting corporations, whose managers believe in nothing but the market *über alles*."[18]

Governmental Constraints on the Media

Besides corporations, advertising agencies, and public relations firms, governments, too, put heavy pressure on the media. In every country governments control the media at least to some extent. This happens in every democratic regime, more so in most socialist countries, but above all in totalitarian regimes and dictatorships.[19] In the United States, the government and business corporations publicly tout the "freedom of the press" as a pillar of American democracy. Both of them, however, require more of the media than the legal and moral demands of avoiding libel and reporting the news in a balanced truthful way.

For instance, the new Telecommunications Act of 1996 promotes a growing concentration of the media, even while it challenges new networks to seize market openings, expands programming opportunities, and increases the control of the survivors. The escalating rate of mergers between corporations and media giants, however, puts increasing pressures on media personnel. Recently the Federal Trade Commission has begun to stir. The White House, Congress, and the FCC have failed to question this grand consolidation of the media.[20].

Governments also have their ways of controlling the sources and quality of information that the media employ. For example, the U.S. military learned an unforgettable lesson from the media's coverage of the Vietnam war. That war, televised with its daily body counts into Americans' living rooms. It also precipitated riots that ended up not only in killing four students at Kent State University, but also in turning public opinion around against continuing the war. Indelibly impressed, the U.S. military changed its policy into censoring news coverage during conflicts. It carried out this policy, beginning with the invasion of Granada and perduring even through the Persian Gulf War. Nor were Margaret Thatcher's ears deaf to this lesson, since she pursued the same policy during the Falkland War.

Again, government offices both flood the media with releases and also shut off access to certain documentary sources. Moreover, the government can issue veiled threats to the media of raising tax burdens or appointing to the FCC persons opposed to the growing influence of the media. In addition, neither government nor big business will allow the media enough freedom to mount a deep persistent critique of them. Yet, when set within the wider context of the American citizenry, this dominative power of business and government has its limits. For, as we saw during the Gulf War, people's tolerance depends on how irritated or alienated, how docile or resistant, they come to feel.

When the Reagan Administration arrogantly refused to accept the verdict of guilty from the World Court in "Nicaragua vs the United States," leaders of the elite media marched in lockstep with the Administration. Nor did those media owners and publishers let their editors dissent from this exclusively self-centered policy of the U.S. government.

Another instance of governmental domination of the media occurred in connection with the United Nation's Law of the Seas. That law aimed more at the good of all nations than at economic interests of some of them. In a response favoring U.S. extraction companies, the State Department took its stand against that proposed law. Accordingly, media owners, publishers, and the elite press permitted very little public criticism of this State Department policy. Cases like these raise many interesting questions, one of which is whether reporters and staff writers or editors and publishers are tethered tighter to those on the higher rungs of the hierarchy above them.

Regarding governmental control, especially since the start of the Reagan Administration, the U.S. government has actually been flooding media offices with a steady flow of releases, all in line with the Administration's view. Every day the State Department holds a press briefing. Meanwhile, not to be outdone, the White House and Pentagon each hold two press briefings daily. Every day between 15 to 20 press releases emerge from the White House, providing plentiful fodder for editors who simply have to have something to tell their audiences--and to do so within tight deadlines. This administrative flood amounts to more than 6,000 releases per year from the White House. To measure its size, one must add to this the more than 1,200 releases per year from the Pentagon and State Department. In addition, these administrative centers create background papers, interviews, leaks, and a wide range of staged events.[21]

Engulfed in that torrent, reporters and editors can hardly perceive public issues in terms other than those gushed forth in this government deluge. What makes matters worse is that, by issuing so many releases, the government sets the parameters for debating crucial issues.[22]

For example, consider how the U.S. government tried to color the way the media pictured Nicaragua in the mid-eighties. After examining government releases on this issue, the Fairness & Accuracy in Reporting organization (FAIR) stated :

> By sheer force of repetition, the administration has driven home its anti-Sandinista propaganda themes in the media. No matter how outrageous the allegation, few reporters bothered to include a simple disclaimer: 'The charge could not be independently verified.'[23]

This is just one instance of the overt side of government control of the media. According to Parenti, the C.I.A. has foreign and domestic journalists on its payroll, hired to convey the Administration's viewpoint to their unsuspecting readers. He also claims that in the early 1950's, the C.I.A. was paying between 400 and 600 journalists for stories of its preference. It secretly commissioned the publication of over 1,200 books. In 1975 the C.I.A. owned outright more than 200 wire services, newspapers, magazines, and 12 book-publishing complexes, even while it subsidized many more. The C.I.A. runs the largest news service in the world with a budget larger than all major wire services put together. And at least 25 news organizations have served the C.I.A., including the *Washington Post*, *New York Times*, *Newsweek, U.S. News and World Report*, and the *Wall Street Journal*, as well as CBS, ABC, NBC, The Associated Press, United Press International, Scripps-Howard newspapers, and Hearst newspapers.[24]

This covert side of media manipulation deserves careful attention. Documents released by the committee investigating the Iran-Contra affair show that Oliver North and Walter Raymond, as members of the National Security Council, steered the State Department's Office of Public Diplomacy, "to orchestrate negative news coverage of Nicaragua."[25] The same documents likewise show that the National Security Agency both instigated leaks of intelligence information and manipulated covert operations in Nicaragua. These were its ways of shaping public opinion in the United States about Nicaragua.

Meanwhile, another agency was developing intricate programs to help the Reagan Administration persuade Congress to renew aid to the Contras.[26] The *Miami Herald* quoted one senior U.S. official as saying, "If you look at it as a whole, the Office of Public Diplomacy was carrying out a huge psychological operation of the kind the military conducts to influence a population in denied or enemy territory."[27] Ironically, people like Oliver North and Robert Owen of a public relations firm called I.D.E.A. Inc., were exchanging among themselves far more truthful perspectives about the Contras and the factual situation in Nicaragua than they released for dissemination to the American public. Moreover, during his frequent flights to a private landing strip in Nicaragua, Robert Owen carried arms to the Contras and gathered information.[28]

A remarkable revelation of governmental efforts to control the press occurred in the Lucia de Cerna case.[29] During the 80's and early 90's, American policy and monies had supported the Salvadoran army in its civil war against revolutionary forces. On the night of November 16, 1989, on the grounds of the University of Central America in El Salvador, soldiers of the Salvadoran army shot to death six Jesuit priests and their two companions. Mrs. de Cerna witnessed this scene and testified before a judge in the Spanish embassy in San Salvador what she had seen. Her testimony exposed the cover-up and brutality of the Salvadoran army as well as the complicity of the U.S. government in this affair. So, the U.S. Administration tried hard to keep Mrs. de Cerna muzzled.

Soon, she was flown to Miami where Salvadoran officers badgered her for hours to change her story. An F.B.I. agent witnessed this psychic abuse without protesting. No agents of the press were allowed to observe these proceedings. Day after day, the Salvadoran officials tried to get Mrs. de Cerna to change her story. They threatened that if she did not do so, they would send her back to El Salvador. There she knew certain death awaited her since she was the sole remaining eye-witness of murders perpetrated by the Salvadoran army. Then, after five days of relentless and unconscionable interrogations, she broke down and denied her story for a while. But once she was released from their psychological torture and had regained her normal composure, she reaffirmed what she had seen.

On hearing of this case, America's major networks and leading newspapers, except the *Boston Globe,* chose to look the other way.[30] For if Americans were led to focus on this case, they would have questioned their government's policy towards El Salvador. Political entente between Reagan and media directors, however, did not permit this.

The U.S. Administration also had other ways of controlling the media and their personnel. Although the Freedom of Information Act had become law, it soon became rather clear that the Administration was devising ways of making it difficult for researchers to access certain kinds of government documents. Updating the procedures it had used in the McCarthy era, the U.S. government kept files on prominent authors and artists for any "pink tints" they might reveal.[31] Working through the Cointelpro system, the Administration refined its computer-based touch on political liberals and radicals.[32] At the National Security Archives, the F.B.I. even tried to enlist librarians as undercover police to report on any suspicious looking persons doing research in the Archives.[33]

The Reagan administration invented many regulatory devices to protect past and present government officials. It constricted the Freedom of Information Act to prevent exposures that might be too revealing. The government enlarged the area of official secret documents. It upgraded documents to "secret" or "top secret." It inserted computer blocks to forestall access to certain entire areas of documents. It injected much bureaucratic red tape in the handling of requests for access to government documents. As a result, legal battles between newspapers and the government focused far less on libel suits and far more on access to fills.[34]

The foregoing has offered some insight into the ways in which big business and the U.S. government control the media. Nevertheless, on their own swing of the pendulum the media fight back. For they often fiercely criticize the government and big business, while exposing or threatening to expose corrupt practices. Yet at the same time, they tremble under threats of losing large advertising accounts and of facing lawsuits from charges of libel or illegal use of government documents.

The Media's Inner Hierarchical Domination

The concluding task of this chapter is to discover how the media's hierarchical system dominates its own agents. We begin at the bottom of the hierarchy. To what does an apprentice writer in the media trade have to submit? How is he or she taught to walk the company line?

A freshly recruited reporter or story writer has to meet the disciplined demands for style, diction, and flair current in the media and in the particular institution where he or she works. This discipline is grueling enough for a young reporter. Yet what frustrates such a person even more is to see one's carefully written report or story gutted by an editorial pen to fit in with the paper's policy. Suppose, for instance, that a careful and talented rookie reporter for the *Los Angeles Times* has drafted a high quality investigative report on schools and drugs in the Watts area. When the rookie finds that some editor has ripped the story apart, he or she becomes angry and discouraged.

Editors sometimes also rap the fingers of more experienced news personnel, as evidenced by the well-known story of Ray Bonner at the *New York Times*. A highly respected expert correspondent on Central America, Bonner had carefully researched a documented exposé on El

Salvador that raised doubts about U.S. policy toward that country. The veteran newsman then submitted his tell-tale account to the *Times* editor. Because of his reputation, the editors decided to publish it but then reassigned him to write for their financial pages. In this way the *Times* effectively silenced him on Central America.[35] Almost every reporter, feature writer, and media person felt the chilling effect of that move.

A reporter or story writer also runs quite quickly into the "bounds of the inexpressible"; that is, those unmentionable topics or places upon which one may not turn the spotlight of public attention. For instance, suppose that an investigative reporter tried in the late 90's to publish a document that opposed the export of harmful pharmaceuticals and pesticides by U.S. corporations to developing countries. In it he would highlight the fact that these drugs and chemicals failed to win minimal approval from the Food and Drug Administration or from the Department of Agriculture. He would build his case on the claim that Americans should not do that harm to others which they refuse to do to themselves. No matter. All such evidence would be regarded as meaningless. For the editors would recognize that this report crossed the "bounds of the inexpressible" and that if they published it, they would feel pressure from the giant interests whose "ox would be gored" by such a report.

Another current "inexpressible" is Anglo-Americans' responsibility to learn Spanish in order to promote civic discourse in the next century. Suppose a bill proposed to establish English as the official language of the U.S. and required new immigrants to pass a minimal competency test in English for citizenship status. Suppose a staff writer recommended the fairness of attaching in reciprocity a certain rider to this proposed bill. This rider would require, out of a sense of shared responsibility for the well-being of our increasingly pluralistic nation, that all English-speaking Americans under 40 years of age pass a minimal proficiency test in Spanish. The staff writer would emphasize that Hispanics will soon be between 20 to 25% of the U.S. population. He or she would stress the benefits if all Americans knew Spanish--it would broaden their minds, facilitate civic discourse, reduce North-South tensions, and open the psychic windows of U. S. citizens to their far more numerous neighbors "south of the border." All these solid reasons would fall on an editor's deaf ears. For the staff writer would have transgressed the "bounds of the inexpressible."

Historically, some human rights lawyers tried to contrast the favoritism shown *La Prensa*, El Salvador's American-sponsored newspaper, and the non-publicity given in the U.S. to El Salvador's two independent newspapers, *El Chronica del Pueblo* and *El Independiente. El Chronica's* editor and photographer were hacked to pieces by death squads. The editor of *El Independiente* fled to Mexico where he was persecuted.[36] In the United States these human rights lawyers, however, soon ran into a brick wall. Media directors would not even answer their question why only *La Prensa* could be wrapped in the colors of a martyred press whereas the other persecuted presses were not to be mentioned. The standard editorial policy was simply that only *La Prensa* was to be portrayed as poor and defenseless. It was to be pictured as standing amid a turbulent sea of chaotic insurrectionist currents, a solitary symbol of embattled freedom, as inspiring as the Statue of Liberty.

Similarly, in the early eighties the U.S. Administration judged Nicaragua an enemy upon "our" hemisphere. America's elite press "loyally" adopted this policy and enforced it.. So, if a reporter did a story on the amazing medical, educational, or social progress achieved by the Nicaraguans after their revolution of 1979, it never saw daylight in major U.S. media because such news was "inexpressible." According to U.S. policy, the Sandinistas simply had as much as possible to be painted as "black sheep" and portrayed as a "truly formidable" enemy of super-power U.S.A.

Nevertheless, most news editors bear only partial blame for such restraints on sound reporting. For they are squeezed just as tightly by their publishers and owners as reporters are by their editors. The iron rule of Arthur Sulzberger, Sr., former executive publisher of the *New York Times,* has become legendary. No one works long in the newspaper business without learning how press magnates like William Randolph Hearst or Henry Robinson Luce dictated the political pressures which their papers were to exert on the public

For instance, in 1949 Hearst noticed a rather obscure preacher named Billy Graham and ordered his editors, "Puff him!" Within months Graham was drawing crowds of over 50,000. The same thing happened with Joe McCarthy. Hearst simply ordered his editors to draw serious attention to that baiter of communists. Even after discovering that McCarthy had fabricated the number of alleged communist sympathizers, Hearst continued his directive.[37]

This control exercised by editors and publishers is felt not only in newspapers, TV, and radio, but also in the book publishing industry as well. In fact, 2% of the nation's publishers control 75% of the books published in the U.S.A. Little wonder, then, that certain viewpoints do not get published easily.[38]

Moreover, as mentioned, the wire services of the Associated Press and the United Press International supply most of the daily non-local news to local media editors. The latter have little to say about how the directors of these news services choose their leading topics nor know how responsible they are internationally. Directors of these wire services never have to account personally in public for the norms by which they select news or emphasize certain approaches.

It also seems puzzling that America's three major TV networks, who compete so hard to win higher Nielsen ratings, do not at times contradict each other in their ways of interpreting the news. If they enjoy as much freedom of the press as they claim, would they not disagree at times in presenting how they read the meaning of events? Yet they do not challenge each other's approaches to news stories. CBS may be a bit more liberal but generally the three TV networks saunter along harmoniously, much like three sailors who, taking shore-leave, stroll in step together along some boardwalk. A consent to appear concordant rather than to risk a chancy harmony seems the networks' factual mode of operating

Then, too, non-concordant views are available to American news offices from papers like the *Economist* in London or *Le Temps* in Paris. Certainly the leading newspapers of Lima and Buenos Aires, or of Lagos and Calcutta represent different voices. Why do not at least a summary of such newspapers' editorial opinions get published regularly in our media? It would make Americans aware of the dialogue of the whole human family? Instead, directors of networks and news services, having already discredited *TASS* and *Pravda*, reenforce Americans' navel gazing by providing sources that arise from the same perspective. By broadcasting such homogenized "news," they "protect" Americans from those legitimate criticisms of the U.S. that arise from the quite different perspectives of other countries and cultures.

In the media hierarchy, then, the owners and directors of the networks and news services, including gargantuan media corporations and such wrestling media magnates as Rupert Murdock and Ted Turner, appear as the real giants who make their power felt by controlling owners, publishers, and editors. Positioned nearly at the top of the media totem pole, they also dominate small-fry story writers and reporters.

Paradoxically, the directors of networks and news services, media owners, and publishers--the persons in the upper echelons of the media hierarchy--clearly grasp one of their duties to the American people. Since they want a continuing large audience and the profit that goes with it, they have to avoid that popular chaos which any heavy erosion of this country's basic belief-system and its basic morals would bring about. Positively, they know they should maintain and enhance the beliefs and values vital to the life of American free society. Yet between this duty and factual performance a great gap indeed can exist.

Until now, we have not examined the way the law controls the media. Beyond reporters, editors, publishers, owners, news services, and networks, the media hierarchy runs upwards to U.S. law. Of course, the law and court precedents have wisely restricted many irresponsible practices of the media. So, media people worth their salt are quite sensitive about being sued for libel. Even apprentices in the media undergo an unrelenting discipline of inserting into their writings such terms as "allegedly," "as it is reported," "according to reliable sources," and so on. But more recently, media attempts to access tell-tale government documents and to defend the anonymity of their sources have generated an ongoing and increasingly sharp battle with the interpreted "law of the land." As a result, in recent years the balance of cases has tended less to libel suits and more to cases involving access to sources.

In the careful study, *Estrangement: America and the World*, Professor Richard Ullman witnessed that the media are dominated more by the State and its policies than the media admit. Back in the '80s, for instance, court interpretations of U.S. law tended to support the desires of a conservative government--desires which the media often followed. The 1996 Telecommunications Act unleashed both competition among media giants and fear among lesser media agencies--fear of being absorbed into still larger corporate hierarchies. Worse still, we currently lack legally established international media-monitoring committees and a U. S. law binding media directors, under stiff fines, to accept the verdicts of such committees. This legal vacuum lets media hierarchs avoid any personal public accounting of how they have directed the media or the news emphases whereby they have led the thinking of the American people. This defective pinnacle, then, of the power-structure dominating the media industry lies in the law and its lacunae.

Summing up our study of media domination thus far, two preliminary steps led to a positive recognition of the media and a clear picture of the media giants. Then we saw how business corporations and government exercise preponderant influence on the media. Lastly, we discovered the ways in which the hierarchs within the media system dominate the various media personnel who are positioned lower down on the totem pole of that system. We have yet to see, in the next chapter, how the media invade the lives of their usual audiences--that is, the ordinary people who are bombarded by media bits and confused by the media maze.

QUESTIONS FOR DISCUSSION

1. In what ways have the media increased your freedom?

2. Comment on Michael Parenti's statement about the American media: ". . .one confronts a precooked, controlled, centralized, national news industry that is in sharp contrast to the 'pluralistic diversity' of opinion that is said to prevail in the U.S."

3. In what ways does big business control the media?

4. How do PR firms and the ad industry dominate the media?

5. In your view, what government agency harnesses the media most? How do you feel about this?

6. What are some of the ways corporate owners and top executives control editors and reporters?

7. How have Third World countries been oppressed through the United States media?

8. If the 1996 Telecommunications Act unleashes more competition in the media industry and more mergers, will this lead to more homogenized or more diversified news service? What evidence supports your answer?

Chapter 8

IMPACT OF TV AND OTHER MEDIA ON THEIR AUDIENCES

The problem of media impact and influence is quite complex and still receiving much study.[1] The term "media-effects" has become a mare's nest in the media sciences. In these sciences, it involves technical distinctions between influences that are strictly causal, or conditional, or contextually reciprocal, or simply statistically co-relational. Then, too, given the psycho-physical nature of media audiences, the strictly causal influences upon them need to be distinguished in terms of physical or psychological, or psycho-physical causality. Accordingly, the present authors recognize the wisdom of William James' pragmatic insistence on eventually immersing one's knowing in the concrete complexity of life as it flows on for ordinary people.[2] In this work, then, the terms "impact" and "influence" will be used at the level of common sense. That is, while respecting the niceties of scientific speech, the authors intend their usage of these terms to carry the meaning which ordinary people give them. So without pretending to speak with the precisions of the media sciences, the authors trust their readers to detect the central thrust and practical consequences which ordinary people detect on hearing the terms "impact" and "influence.

This chapter, then, examines both the positive and the less helpful impacts and influences of the media, especially of TV, on their various kinds of audiences. The commitment was made in Chapter 2 to maintain a primarily positive approach to the media. The present exploration of the various impacts of the media will test that commitment profoundly. For the present chapter must <u>balance</u> two things. While recognizing the negative influences of the media, it must insist on the media's positive achievements and potential for greater service of the human family.

Chapter 7 showed how business and government put pressure on the media and how powerful hierarchs within the media control media members below them. The present chapter shifts the question to ask: How do the media affect their receivers' thought, values and behavior? In other words, it focuses on how TV and the other media influence their audiences' minds, hearts and deeds. Since the media exercise powerful initiatives of their own and yet also respond in part to their audiences' needs and wants, to their sensitivities and permissiveness, the media are caught in a bi-polar, to-and-fro dynamic as they interact with their audiences. Accordingly, as a starter, some notice of audience impact on the media is in order. Admittedly, the media influence their receivers' thoughts, values, and behavior far more than their audiences influence the media. Nevertheless, some recognition of this real, though far weaker, "flip side" of the question of media influence deserves notice.

Audiences influence the media by being too passive, uncritical, and unprotesting or imitating the Roman mob of old by crying out for "bread and circuses."[3] As a result, audiences frequently receive from the media a lower level of content that they rightfully and legally deserve, namely, a content that genuinely promotes their development as human persons. Not getting the content they deserve, audiences unfortunately get from the media what they passively influence the media to broadcast to them, or even worse, the pap and pulp they passionately call the media to supply. Some instances of this audience influence on the media support this claim.

In general, newspapers, radio, and TV cater to people's demands for local news or news acceptable to their local or regional audience. Local newspapers and TV highlight local murders, problems at city hall, and the way local stocks are faring currently.. For example, in the New York City area the heavily Jewish population calls the *New York Times* to present a strongly pro-Israel position. This prevents equal coverage of Israeli atrocities against the many Arabs locked in Israeli prisons or of Israeli retaliatory air-strikes that leave Arab women and children slaughtered. If the *Times* gave balanced coverage to this conflict, their circulation, as well as their ad accounts, would plummet.

Again, look what changed President Johnson's thinking about the Vietnam war. Once he saw Walter Cronkite on TV first tipping his hand against that war, Johnson knew that public sentiment against the war had become too strong to buck. Johnson knew that CBS reflected the people's widespread repugnance to continuing the war and that Cronkite was simply serving as the mouthpiece of middle America.[4]

On the other hand, and as our principal concern here, the media exert a preponderant influence on the values, thoughts, and even behaviors of their audience. This is especially true of TV which impresses its views upon the eyes, ears, and minds of those watching it. With its fascinating appeal, TV lures the average American to view its screen for about three or four hours daily and affects its viewers' minds much more profoundly than other less frequented media. Media critic, Neil Postman, describes in full these mental effects, which are here condensed into a bare general listing. His evidence shows that TV usually leaves people unrefreshed, supplies misleading bits of information, devastates political discourse in America, anesthetizes viewers to a sense of history, and fosters a phoney sense of religion as entertainment and of education as amusement.[5] Such a list, if not counterbalanced by TV's achievements and potentials, could lead to a primarily negative attitude toward TV and even perhaps toward the media in general.

Writing specifically about newscasts, Gregor Goethals supplies additional reason for caution:

> The nightly news programs, for example, present accounts of "reality," carefully constructed narratives made up of both words and images that explain what is happening in the world. The success of television's ritualistic orientation may be measured by the fact that most of us do not view the "news" as a symbolic construction, but as an essentially believable reality.[6]

That is, audiences usually take TV news too gullibly, as the whole "believable reality" instead of as a highly selective art form aimed at tilting minds in one particular direction.

To examine how TV and other media impact the minds of their audiences, we here select a different tone and path from Postman's. The foci of this chapter are how TV and the other media: 1) implant values or fixations, 2) expand or narrow the minds and hearts of their audiences, 3) exploit violence and sex, 4) lure to rash judgment and snap decisions, and 5) promote consumerism. These five headings provide a map for the present chapter, while the last three headings also invite suggestion for offsetting these negative impacts. We close with an American-based test to measure three ways of audience response to the media's influence.

Media Implant Values or Fixations in the Minds of their Audiences

The media generally, and TV especially, deserve gold medals for their constant emphasis on certain values that humanize people. Good health, freedom, and democracy form a trio of such values which the media continually reinforce.

The media promptly inform their audiences of the most recent medical breakthroughs and of the dangers of spreading epidemics. For instance, although media exposure to movies and sit-coms that dealt with AIDS correlated with increased judgments of personal risk, exposure to news programs covering AIDS seems to have decreased viewers' perceptions of personal risk.[7] By warnings about storms, flooding and treacherous road conditions, they protect thousands, sometimes millions, of lives. Even commercials about remedies for obesity or about opportunities for various forms of exercise keep stress-prone Americans aware of the need of efforts to maintain their health.

The media's so obvious role in promoting freedom often evokes criticism that most media are "too liberal." Yet to weigh its worth, this promotion of freedom needs explicit attention. TV presentations of Shakespeare or of historical documentaries highlight the role of freedom in their leading characters--say in Macbeth or Hamlet, in Lee at Gettysburg or Churchill in the Battle for Britain. TV press conferences with the President or his spokesperson demonstrate concretely how one kind of freedom operates. Commercials in general and political commercials in particular appeal to the freedom of buyers or voters. So much does TV emphasize individual freedom that it has reinforced viewers' right to backfire by voting against political candidates whose commercials attack their opponents' character.[8] Moreover, TV has led viewers to switch channels freely during commercials, especially by using remote control. For instance, in a test study, 40% of TV viewers did no zapping, but the remaining 60% zapped to avoid most commercials, if not all.[9]

Democracy is surely one of the media's most cherished and promoted ideals. Studies document the role of TV and radio in promoting liberation movements that call for democratic governments--such as those that occurred in the 50's and 60's in former African colonies,[10] or in '89 during the collapse of communism in the Soviet Union and its former satellites.

Besides impressing these positive ideals in their viewers' minds, TV, radio, and other media have also uncritically impressed some negative themes upon Americans' thinking. By reinforcing certain stereotypes and labels, the media influence the minds of their audience. Consider three common instances: anti-communism, and--now that Soviet communism has collapsed--negative attitudes toward developing countries, and finally, the various slogans for conveying the theme of "America, number one! "

Concerning the stereotype of anti-communism, the media caught that virus, which culminated during the Cold War, from the Administration's policy of containing the enemy. As a result and by constant repetition, the media instilled into the American people a deep fear of Communism. Starting from the days of Joseph McCarthy and throughout the Cold War, the media have reinforced this knee-jerk anti-communist reaction ever more deeply, both here and abroad. Nor has the end of the Cold War uprooted this media-sown fear. One need only say "Marxist" or "radical" to a random sampling of fifty Americans to tally up how many negative reactions still occur.

Today, after the fall of Soviet communism, a major problem for media directors is unfortunately, not whether we need an enemy, but how to find one. Through its anti-communist propaganda, the media have so deeply ingrained the need for an enemy that many Americans can scarcely perceive their identity--who they are--except as a people threatened by an enemy. As we shall see, the current vacuum is largely being filled with a fear of the world's vast majority of its impoverished and oppressed peoples.

The media slip into a second way of needlessly fixing a theme upon their audience when they refer negatively to peoples in less developed countries than the U.S. Forms of this stereotype surface, for instance, in images of Arabs as "shifty," of Latins as "lazy," of Chinese as "closed-in and crafty," and of African groups as "rowdy rabble." By preferring such images, the Western media further entrap the peoples of the poor nations and Westerners themselves.[11] Anthony Smith, a political scientist at Oxford, holds that even before the communist party in the U.S.S.R. dissolved itself, the American media had already chosen a new punching bag, the so-called Third World.[12] Although the term "Third World" is arrogant, inaccurate and obsolete, it still is widely used. It is arrogant since it makes the West a First World or "number one." It is inaccurate because it in fact points to the two-thirds of the world's population which is economically poor by U.S. standards. It is also obsolete because communist socialism, the so-called Second World, has, except for places like China, left the world stage.

If one focuses on Cairo, Egypt, as the center of the earth's globe and sees the United States, Europe and Japan as the resultant northwest quadrant of that globe, the picture becomes clear. Media in that northwest quadrant are already saddled with a pre-existing negative image of the economically poor nations of the world. Unfortunately, these media reproduce and broadcast this negative image. These media are fashioning in the minds of their audiences a negative association about the world's economically poor peoples just as effectively as they earlier forged the anti-communist bias.

Since decolonization in the 50's and 60's, the economically poor countries of the world have for good reasons formed a negative estimate of these North westerners (from Europe, U.S.A., and Japan). Peoples from less developed countries know historically that many such North-westerners once viewed themselves as duty-bound to carry the colored peoples of the world as "the white man's burden"--an attitude that was at least demeaning of the Southerners. Moreover, these peoples know quite well that the Northwesterners have been oppressive colonizers and the authors of many inept development projects that have failed.

If they chose, northwestern media could emphasize the opportunity for mutual enrichment that emerges from creatively appreciating the diversity of different nations. They could also challenge Americans to dialogue fairly with these peoples whose cultures often are far older and wiser than those of the globe's northwestern quadrant, particularly than the culture of the United States.[13]

Anthony Smith has shown the invisible captivity to which the electronic media of this northwest quadrant are subjecting two-thirds of the world's population. For instance, he writes:

> To be imprisoned inside the misinterpretation and misunderstanding of others can be a withering form of incarceration. It is a fate which can afflict whole nations and cultures as painfully as individuals. Today, two-thirds of the world's population is locked into a quickening spiral of deprivation, while committed avowedly to a goal known as 'development.' The new insistence on the part of Third World countries that some kind of restructuring must take place in the machinery of international communication is one part of their struggle to gain control of the processes of their economies.[14]

This explains why so many nations want a new deal in the media. Smith states their reasons for this quite clearly:

> The Third World has accused the West of cultural domination through its control of the major news-collecting resources of the world, through the unstinted flow of its cultural products across the world, and through the financial power of its advertising agencies, its international newspaper chains, its news-print companies and its hold over the electro-magnetic spectrum on which broadcasting, navigation, meteorology, and much else depend. The swamping effect of this vast machinery has transforrned the social fabric of Third World countries as it has repressed its traditional cultures.[15]

The stereotype of anti-Third-Worldism narrows the viewpoint of American audiences at a time when the internationalization of consciousness is urgently needed. The 21st century rapidly approaches and all Americans are urgently called by the rest of the world to become more aware of world-wide needs and concerns. Our only earth is direly threatened with ecological collapse, population explosion, nuclear proliferation, and terrorist blackmail. These disasters can only be avoided if the media lead people to think globally. Only with such thinking will people learn to discipline themselves both to a reasonable restraint in using resources and to a humane cooperation with peoples of quite different cultures.

The theme, "America, number one!" supplies a third instance of the media's fixing a stereotypical slogan upon the minds of Americans. First, the facts need noticing, then a reasonable response to them.

Militarily and economically, of course, America has been and currently is number one. In the 80's, the media whipped up support for the Administration by touting Star Wars, by displaying nuclear submarines and Stealth bombers, and by glorifying our military smashing into Panama, Libya, and Iraq. With the collapse of Communism and some downsizing of defense in the 90's, the media have somewhat reduced their heavily militaristic tone and broadcasts. Yet before the Clinton re-election of 1996, they again glorified a repeated U.S. squashing of southern Iraq in defense of one group of Kurds in northern Iraq. They glamorized the flying in of cruise missiles from distant Guam--usually without telling taxpayers that this transport alone cost them two million dollars.[16] Moreover, while the media keep praising our economic prowess, openings to new markets, and still more powerful corporate mergers, the U.S. military still consumes a giant share of the annual federal budget.

It would have tarnished the media's stereotype of "America, number one!" if during Desert Storm they had balanced their triumphalistic accounts with factual reports of the killed, wounded, and starving Iraqi soldiers and civilians. For usually, when news coverage emphasizes military muscle-flexing, any call to face those facts which undermine the dominant and self-serving image irritates most Americans. Through their usual excessively nationalistic coverage, the American media have schooled their audiences to the automatic association that American soldiers are bigger, stronger, and more important than foreign soldiers. Media managers often judge it unwise and perhaps unpatriotic to broadcast facts counter-indicative of "America, number one!" So they avoid balancing our strengths with our weaknesses and inculcating a modest national self-esteem.

Were they to do so, they might quip that America is also number one, not only in armor and money, but also in debt and pollution. That, however, is not a laughing matter to be quipped about. When pointing out that America has lost much of its financial and industrial leadership to Japan and the European Union, the media might broadcast the predictable results of these shifts. Already the media inform us that America is not number one in providing jobs, health care, and affordable housing, in reducing crime, violence, drug usage, and infant mortality, or in achieving high scores in schooling, racial harmony, and women's rights. But in doing so, they usually do not place Americans within the larger context of the embarrassing statistics from European and other nations on these points. To add this context, however, would foster a modest national self-esteem It would also help free media audiences from the stereotype "America, number one!"

In view of these many negativities, however, media directors have some seeming reasons for maintaining this stereotype. For otherwise, how support Americans' self-esteem and keep them from discouragement in the face of all their problems? On the other hand, an unbalanced and misleading message can engender neither sound self-esteem nor the courage Americans need. All the many ways of sending the message "America, number one!" only appeal to the erring spirits of anti-social independence and excessive individualism.[17] Furthermore, through the U.S. media's arrogant self-vaunting of America, this widely broadcast stereotype irritates the vast majority of the world's population and isolates Americans still more from the community of nations.

For media directors some remedy to this dilemma may lie in their stressing how Americans are great in other ways than mainly in their military and economic might. In brief, they can emphasize how many kinds of freedoms Americans enjoy and show episodes about these freedoms--often unknown or undreamed of in many parts of the world. With appropriate analysis, U.S. media can broadcast how exceptionally committed to work most Americans are, compared to many peoples in less developed countries. On this theme they can supply episodes both to show what made America great and to hearten America's large majority who are workers. Even more than at present, they can highlight America's ascendancy in science, technology, and the arts. To these thematic presentations, episodic illustrations can be inserted showing that these breakthroughs uplift nations more widely and permanently than does opening a McDonalds in Moscow or getting a Coca Cola franchise in Malawi.

Moreover, in both domestic and foreign broadcasts, U.S. media could wisely emphasize the strong religious attitudes and practices of most Americans. This trait distinguishes them from many other nations. In various episodes the media can reveal those energies of trust in God and hope in his help that have given strong embodiment to a key American assertion, "one nation under God." Let movies and TV artfully show that the deeply religious faith of most Americans has been strong enough both to acknowledge their many shortcomings and to trust the Holy One to "heal our every flaw." In such ways as these, then, the media could modulate the sense of "America, number one" so that Americans may become more modest, diversified, and realistic, and not be viewed abroad primarily as a military or economic bully.

If the media freed themselves and their audiences from these negative fixations, they would exert a far more beneficial influence. To do so they can invite their audiences to emphasize cooperation and social skills more than competition and isolated individualism. In this way the media can create an audience of more humanized people, especially in the younger viewers of TV.

Expanding or Narrowing Hearts and Minds

Positively, TV and the other media entertain people tired out from work and stressed out by their problems. In America's heavily workaholic culture, the media provide a relaxing milieu of play and amusement. Their efficacy as stress-relievers has been scientifically examined.[18] For instance, it was found that stressed women watched more TV, especially more game and variety shows. Stressed men viewed more programs featuring action and violence, e.g. sports and crime series.

The media also expand the minds and hearts of their audiences by their globalizing and socializing influences. When millions of TV viewers first beheld that scene telecast from a spaceship of the blue planet floating alone in black space, a consciousness of "our only earth" arose among many of them. Similarly, when viewers saw Neil Armstrong take man's first step on the moon, or delighted with those Germans dancing atop a crumpling Berlin Wall, or felt the weary feet of Rwandan refugees tramping homeward after years of miserable life in ramshackle refugee camps, the power of TV to globalize human consciousness became evident. It powerfully created a sense of achievement, solidarity, and compassion with all human beings caught in such hopes and fears, in such sufferings and joys.

Besides effecting such expansions of mind and heart, TV and the other media sometimes narrow those same minds and hearts. They can so emphasize entertainment that Americans find themselves "amused to death," as Neil Postman puts it. Generally, prolonged viewing of TV not only inhibits creative thinking in children and adults, but also forms an addiction in more than a few instances.[19] In many cases, TV viewing fosters an excessive individualism and an oblivion to others needs. For example, one viewer's choice often leaves the interests of other co-viewers unattended. Moreover, viewing distant people on TV often results in a less positive perception of them than if they were met face to face.[20] These are some of TV's narrowing effects on their audience to which more notice needs be given.

Although some media magnates regard it as an obsolete formality, the Communication Act of 1934--a mandate reinforced in 1974-- still requires all media directors to serve the "public interest, convenience and necessity" especially in regard to children.[21] Since 1934, the necessity of internationalizing the consciousness of adults, and especially of children, has grown exponentially. Because Americans number only 5% of the earth's population, the media need, for truth's sake, to widen their focus away from concentrating mainly on our national interests. They need to stress much more the global interests of the whole human family as well as of Americans. All audiences need an intensified sense and awareness of international concerns. Only in this way will the U.S. media invest effective efforts toward helping heal our much damaged earth and toward gathering together diverse cultures into a unified cooperative human community of nations.

It is especially U.S. movies and TV that teach models of behavior to Americans and media viewers worldwide who watch U.S. movies.

They do so by presenting models that are often artificial and stifling. As Kenneth J. Gergen describes it:

> . . . sincerity is also bled from the common, informal modes of cultural expression. The media expose society to a massive array of self-representations. Our manners of friendship, family relationships, romance, and animosity are documented, scrutinized, rhapsodized and satirized. We see how the lips are pursed in moments of quiet anger, the fingers move to the mouth as expressions of serious thought, and the tongue flicks backward in moments of cautious reflections. As these images are exposed increasingly to the culture, they become the standards for expressions: a subtle Miss Manners for the world of informal relations. They inform the culture, for example, as to how sadness is done--its durations, its modes of expressions, and its proper intensity on various occasions. Indeed in failing to meet these commonly recognized standards, one cannot properly define oneself as "being sad"--either to others or oneself.[22]

Besides teaching this constriction of social behavior, TV and the movies also focus childrens' minds away from reading and from cultivating an early habit of reading. The Yale Family Television Research and Consultation Center has shown that children who watch TV the most are the ones whose reading levels, readiness to read, imaginativeness, and complexity of language use are all lower than normal. Meanwhile, those children scored highest on reading recognition tests in homes where parents mediated TV-viewing most and the children's TV viewing was least frequent.[23]

If only through vividness of impression and length of exposure to media, leaders in the field of media have more psychological impact on their audiences than do leaders in schools and churches. Yet strangely, even when they broadcast episodes from around the world, they rarely address global themes and issues in depth. They do not adequately serve the people's necessity of becoming more internationally minded. Media managers do not seem to shoulder their share of this urgent task to promote a deeper operative sense of Americans' membership in the one human family. In proportion as their contact with audiences is more intense and extended, media managers have a larger share in this task.

Another way the media constrict their audiences' perspective lies in their distracting their audiences away from genuine issues. Media policy avoids acknowledging the basic good sense in most people who want to face reality and tussle with the deep challenges confronting them. Media, business, and government have learned one lesson from Vietnam--what results when the American public find out about My Lai, napalm bombing of innocents, and similar national immoralities. Although some TV movies have recently focused on problems of child abuse, drugs, co-dependency and poverty, TV still remains an entertainment industry before all else. So TV amuses Americans to death and overall, by its general run of sit-coms, soaps, and quiz shows distracts their attention away from problems which, if once faced might disturb the status quo.[24]

For instance, the media focus the attention of tens of millions of Americans upon sport spectacles--both college and professional football, the world series, Olympic games, and other sports events. They zero in on civic, church, and media spectaculars--trips of the President, staged victory parades, papal visits, Oscar and Emmy awards, and Miss America contests. If these distractions fail, then the ordinary "soaps," game shows and crime stories will keep the American mind at a superficial level. So effective is this intent to distract by mere entertainment and amusement that relatively small audiences turn to programs discussing significant issues. Few tune in to Jim Lehrer with his discussion of current problems, or to Bill Moyers with his emphasis on alternative medicines, or central themes found in *Genesis*. Does this shortcoming derive mainly from the fatigue of media audiences or from media directors' failure to live up to their educative role of raising standards for living in a fuller, more excellent way? Otherwise, for what besides profit-making are these directors using their awesome power? In general, then, TV and the other media quite effectively distract Americans from significant issues lest, once aroused, they begin rocking the boat.

Media leaders at local levels are responsible for avoiding pandering to exclusively local interests. They need to challenge the narrow-mindedness of local folk to recognize that the peoples of our earth's other countries are just as real as themselves and that we have common problems. The global family is currently confronted with pressing problems of immense weight and challenged to a cooperative solution of them. Only with a fully global consciousness will people have enough of the courageous patience and creative insight to fashion humankind's great community. The preponderant influence which the media exert make them the premier artists in this fashioning. Hence it seems fitting in our study of the three remaining areas of the media's influence upon their audiences to offer suggestions to these artists for offsetting negative impacts.

Offsetting Violence and Irresponsible Sex by Using a Different Grade of Sources

By portraying violence and sexual irresponsibility, the media, and especially TV, can undoubtedly grab audience attention and thus promote their ratings and their dependent ad accounts. Many studies document the media's excessive emphasis on violence, irresponsible sex, and gender stereotypes.[25] We avoid entering the disputed cause-effect debate between viewing TV violence and aggressive behavior. We also recognize that normal human life involves cooperation and competition, using peaceful industry, yet sometimes also violence, to wrest food from nature or defend family and neighbors. Hence, when we speak of excessive violence in TV and movies, we mean such unusually frequent scenes of violence as tip the viewers to act impulsively and aggressively more readily than act as patient peace builders. An effective curb to these excesses lies in the power of the media to nurture in their audiences a sense of the precious, mysterious, and holy, especially as this sense operates in response to the human body. Media magnates profess a commitment to respond to the pluralism that America features and fosters. Let them then often broadcast those significant excerpts from Eastern and Western religions' traditional wisdoms--excerpts of the basic values these religions witness to. Thus media directors need to escape from the monotony of usually voicing a secularistic mind set by frequently broadcasting the religious mind set typical of most Americans. The secularistic mind set ends up bankrupt when it comes to effectively resisting violence and irresponsible sex . Only a deep intouchness with a human person's spirituality, which relates one faith, religion, and the Holy, can supply the needed inner restraint and the fitting curb to mindless lusting and violence. Accordingly, more sharply focused attention to the way TV and the other media promote violence, sexual irresponsibility , and gender stereotyping seems in order.

The results of a sampling carried out by the Center for Media and Public Affairs in Washington, D.C., prove instructive. For in that city the Center taped 180 hours of programming, as found in a single day on 10 channels from 6 a.m. until midnight These were the affiliates of ABC, CBS, NBC, Fox and PBS, one non-affiliated station, plus the cable channels WTBS, the USA Network, MTV, and HBO. The program schedules were notable only for their ordinariness, and carried no movies of unusual violence. The tapes were then tabulated, computerized and analyzed. The results of those 180 hours of programming are as follows:

1) 1,846 acts of violence--purposeful, overt, deliberate behavior involving physical force against other individuals;
2) 175 scenes in which violence resulted in one or more fatalities;
3) 389 scenes depicting serious assaults;
4) 362 scenes involving gunplay;
5) 673 depictions of punching, pushing, slapping, dragging and other physically hostile acts; and
6) 226 scenes of menacing threats with a weapon.[26]

It may be difficult to demonstrate scientifically a causal link between numerous scenes of TV violence and lived-out aggressive behavior, but such scenes surely do not directly promote personal, familial, and civic peace and harmony.

America has grown from one of the safest countries on earth in the mid-fifties to one of the most violent in the mid-nineties. We do not go so far as to assert a direct linkage of a trend to violence with TV. Yet the viewing of such scenes of violence, especially by young men between ages 15 and 30, suggest a high degree of correlation. In 1953, the Detroit murder total was 130, with 321 in New York and 82 in Los Angeles. Thirty years later the Detroit murder tally was up to 726, the New York toll 1,665, and the Los Angeles murder total was 1,126.[27] Self-defensively, one might focus on the slight decrease in the rate of murders in the mid-nineties and miss the overall 400% to 600% rise in that rate during the past forty years.

Moreover, the sexual behaviors and attitudes broadcast by TV and the movies often induce irresponsible sexual attitudes and behaviors in viewers. Music videos are equally at fault. For researchers Barry L. Sherman and Joseph R. Dominick have found that about a quarter of all characters in music videos are presented in suggestive or provocative dress. They also found that on the average a music video displays four sexual activities, while its emphasis is on titillation and physical activity of a very adolescent nature, devoid of emotional involvement.[28] In the real world sex is, in part, a fun-game. Yet leaving this factual basis, videos, movies, and many TV shows move into a fantasy world where sex is portrayed as *chiefly* a fun-game without emphasizing the gentle respect needed in intimate interpersonal relations or hinting at the consequences of promiscuity. Moreover, the movies and TV have the general power to liberate women from gender stereotypes and have moved in this direction since the 70's. Yet one may wonder whether the reader has discovered many recent videos, movies, or TV shows that portrayed the respect and reverence due the body of every living human being?

For instance, amid the mid-90's concern with the growing AIDS epidemic and teen-age pregnancies, the movie *The People vs Larry Flint* appeared. This movie might have served the common good by emphasizing the urgent need to treasure one's own and others' bodies enough to discipline and refrain from the lusts that lead to AIDS, other forms of VD, and irresponsible sex. Instead, this movie glamorized Flint, a debauched hustler of sleaze magazines, and glorified him and his gospel of voyeurism and promiscuous sex. Moreover, in the end, by using a clever and expensive lawyer, a handicapped Flint is acquitted of libel through the Supreme Court's reading of the First Amendment's protection of free speech. Hereupon, in hero worship, Hollywood wraps Flint in the red, white and blue of an American flag and the crescendo of patriotic music.

Moreover, this movie portrayed women primarily as sex-objects, as kittens for sexual play. Members of the feminist movement rightly protest strongly against such stereotypes. Unlike this film, American movies and TV in the 90's have generally been portraying women with more respect and equality than in the 70's, for instance in the TV serial *Lois and Clark.*[29] Yet if gender stereotyping has been decreasing in main line TV shows and movies, the trend to extend patterns of gender stereotyping into certain videocassettes and cable TV shows has not decreased.[30]

The experience of a normal healthy couple, married for six years, witnesses to the media's noxious influence in this area. Since grade school this couple have seen shows where a man and woman relate with sexual intimacy. Now, however, after six years of marital experience, they feel almost nauseated when encountering broadcasts of bedroom scenes wherein sexual relations are cheapened by an off-hand casualness and lack of respect. Without articulating it, this couple has learned from real life that the error constantly engendered by the media is that sexual intercourse equals genuine interpersonal intimacy. They know many friends who have accepted this misleading and destructive equation and then ended up wrecking their interpersonal relations either before or during marriage.

This disillusioned couple also finds quite unhelpful those TV or movie scenes of casual sex portrayed as genuine intimacy. When they were newlyweds, these scenes hindered them in the adjustments they needed during their first months of married life. They also suspect that much older couples are often misled by these romanticized bed-scenes.

This couple feel that the ideal of sexual prowess propagated by the media must keep disappointing most work-worn couples. For the awareness of their factual performances in bed keeps contrasting too sharply against the unattainable myth pictured by the media. This couple senses the media's repulsive manipulation of millions by using vivid sex scenes, often joined with violence, to grab audience attention.

Offset the Lure to Rash Judgment and Snap Decisions by Emphasizing Wisdom-Models

Artfully created TV programs, along with interesting biographies and sane recordings, can nourish habits of wise decision-making. For such programs display sages in action and create moods where the going order of the day is to foresee a range of possible courses of action and think out their consequences. Can adolescents imitate Palumbo as he pauses to think and ask his seemingly stupid questions? Can high schoolers and collegians watch the crew of spaceship *Enterprise* striving with its Captain--whether man or woman--in doing cooperative thinking before coming to a decision? Before their eyes these viewers of *Startrek* see what it means to gather facts, get counsel, and weigh the pros and cons of this or that possible course of action before coming to a decision. In a program like *Murder She Wrote*, can adults carefully notice how seemingly trivial items become possible clues <u>if</u> one traces out how they are linked with other events and the statements of certain participants? Wholesome TV series, movies, and books, then, can train their young and adult audiences to think more carefully and sanely. And the healthy elderly in retirement homes can select high level TV programs to seed their intelligent conversations and promote their personal growth.[31]

Yet one all-too-pervasive result of TV shows is to foster in their viewers habits of quick associations, rash judgments, and snap decisions. So, some attention to this result is needed to assist viewers to counteract this negative impact..

To spend many hours viewing TV and movies is to see many of life's toughest problems solved in an hour and a half or even in a half-hour. With only twenty-two minutes left after ads, TV sit-coms, detective shows, and newscasts have only enough time for the barest clues to steer their audiences to guess at the underlying motives of the actors. Such programming trains viewers in habits of quickly condemning a person or blackballing a whole group on very slim evidence. Schooled by the fictional life shown in the media, people become trained to leap from a single clue to a rash judgment in real life.

As a medium, TV is neither good or bad and its effects depend on the types of programs chosen and the way in which viewers use them. Yet children's unsupervised use of TV and video games exposes them to interferences in learning, training in impulsive decision making, and still worse effects. One recent study examined the role of background media while teenagers did their homework. It found that, in general, students found that their performance on learning assignments, (in contrast to mere paper-and-pencil assignments) was impeded by the use of background media such as music or a TV program.[32] Furthermore, the powerful pressure of sophisticated advertisements, especially for toys at Christmas time, pushes children to the snap decisions of simply seeing and wanting. So, too, do their playing adventurous video games where the way to win lies in quick impulsive decision-making, rather than in gathering more facts and weighing several lines of evidence. Studies show that the deleterious or healthy cognitive habits which children develop from interacting with TV or video games correlate directly with the absence or presence of family patterns of viewing. And the latter correlate with the social institutions and conditions in which the families live.[33] Finally, children's' unsupervised viewing of TV shows raises concern about their viewing and perhaps being led into interactive erotica.[34]

The increasingly fast pace of many TV shows and commercials also primes people to want instant solutions, to hate waiting, and so to make snap decisions. Here the whole electronic age, as well as TV and movies, bears the responsibility. In the "good old days" before e-mail and faxes, complaints and protests were carried by U.S. mail. Before drafting a response by mail people had time to ponder a thoughtful reply. Not so anymore. Many users of e-mail and faxes want instant replies, given in a tit-for-tat way that imitates the fast moving interaction of half-hour sit-coms. Clearly, such a frenzied pace of communication hardly promotes the humane dialogue that takes pains to uncover the peculiarities of individual situations in order to respond fittingly. The temptation to fall into the quick-reaction mode surely attends our electronic age.

Offset Consumerism by Emphasizing Help to the Needy

At times of natural or man-made tragedies, TV and other media have shown scenes of the various victims of hurricanes, war, crime, AIDS, and starvation. They have thus often elicited compassion for the needy and energized help campaigns on a national, or even global scale. In this way these media counterbalance their almost unremitting trend to foster ever more consumerism. This trend needs both in-depth study and strong counter-action if people are to become more human rather than being captivated by ever more buying and having.

The direct and indirect alliance of the media with business inculcates the consumerist mind set. Granted that in a capitalist society commercials are the price necessary to make possible for-profit TV and newspapers. Still, beyond the cost added by advertising to each product purchased, Americans also pay for advertising by being brainwashed into the consumerist mind set, that is, by being mentally trained to want more and more things.

For example, if a person considers the effects of TV on the inhabitants of poor dwellings in an inner-city slum, some telling questions will arise. How many TV sets there have lured their viewers-- poor whites, Afro-Americans, and Hispanics--into believing, each day more and more, that stylish cars, fancy furs, and glitzy parlors are the hallmark of success? Television has also led many of these impoverished people to believe that before one gets stepped on, one has to step on others if one wants to climb up to the monied peak.

Not upon inner-city dwellers alone is this media power concentrated, but rather, by promoting consumerism through every technique of advertising, this power changes the consciousness of far more people just as profoundly. For few Americans can claim to have remained untouched by media advertising. Through their intensely persuasive hype, commercials steadily lure hundreds of millions of TV viewers, radio listeners, and newspaper readers to buy ever more. Besides being explicitly influenced toward consumerism by commercials, the same millions of potential purchasers are continually being bombarded by the come-ons hidden within the greed-glamorizing ambiance of glittering lotteries, informercials, and ordinary sit-coms. In addition, most Americans have also accepted the media's view of how to climb to the top, even though the many frail and lonely who have won success this way have indeed paid a heavy price. With their various presentations of glitz and glamor, the media have also been the key persuaders who keep Americans' imaginations and dreams focused on consuming ever more and climbing to the top of the ladder, no matter what. So, when a person turns to public broadcasting to watch a show without being interrupted by commercials, he or she feels relief and a welcome freedom from bombardment by advertising.

In sum, we have seen that the media are working profound changes in their audiences' attitudes and ways of thinking. They make them eager for the next impression, rather than intent on probing the meaning of what they are experiencing. They draw most people into a mood of being entertained rather than of doing critical thinking. They incline audiences to focus more on appearance than reality, to be more concerned about the way a person looks than about the way that person is. Television and movies school us to value personality more than the person, the pleasing smile more than a morally honest intent, the politically correct phrase more than the full, perhaps partly embarrassing truth.

After examining many instances of Americans being less than truthful to themselves and others, the well-known author, Scott Peck, identified Americans generally as the "People of the Lie."[35] When people shrug off this accusation, they show both that they no longer sense their own self-deception, and that they are no longer interested in becoming Americans who live and speak like "honest Abe." As a result, America's younger generation--those born after 1975--have in fact become psychically saturated with the media's countless scenes and accounts of deceit, corruption, scandal, and fraud. Many American youths view their world with suspicion and skepticism, and adopt a "who cares?" attitude. Surely America's older generations, and media directors especially, bear more than a little responsibility for this soured reserve and cynicism.

Along with sensationalist papers and fiction paperbacks, many movies and TV programs tend to "dumb down" the popular masses. This reinforces the negative self-image that psychologists find so widespread. Yet precisely when has America more urgently needed better informed and more independent thinkers? When has the need been greater for citizens who can publicly exercise rationally grounded criticism of suggested or enacted policies? A genuinely vigorous democracy does not, after all, sprout up from a crop of couch potatoes. One way to test the effect of the media upon the masses they reach is to select the criteria of a classical American philosopher of community, Josiah Royce. In 1913, he identified three basic attitudes of will as fundamental criteria of ethical health or illness. According to Royce, ethical health is found only where the third criterion operates. These criteria are:

1. an individualistic self-assertion,
2. an individualistic self-withdrawal from the world, and
3. a lovingly loyal commitment to, and dedicated service of both one's closest community and the Universal Community of all minded beings.[36]

Overall, then, which of these three attitudes of will has the media fostered more in the psyches of Americans?

Regarding this first attitude of will, sociologist Robert Bellah and his fellow-researchers have documented the growth of what seems to be a cancerous individualism in America.[37] Of this diseased form of individualism, the mass media are arguably the most powerful inculcators. Even though people with this attitude of will may still mean to care, in practice this kind of individualism makes them have little consideration for others and less for their responsibilities to communities.

For example, take the simple case of the average TV viewer's use of the remote control device. The grandfather of American psychology, William James, taught that whether one wishes it or not, each choice a person makes is indelibly inscribed in his or her brain. Granted then, that whether a person chooses to read a book or take a walk or change a TV channel, that choice is inscribed in the person's brain and thereby he or she becomes at least a bit more like that choice. Moreover, while every organism selects what is in its best interest, every organism also has a sense of its community which calls for some entry into its choices. This holds true for a herd of caribou, a school of fish, a bevy of quail, and hopefully for the human members of a family or nation. Yet in human beings this sense of community can become dulled.

Notice, then, the thousands of a-social choices made by a person using the remote control device before a TV set during a single month. Such a TV-viewer decides: I want this channel, neither that one, nor that other. I want more volume, not less. I want to skip to another channel. I want to surf channels, skimming off from each what's most interesting for me. I want to watch these shows so I won't have to do homework, or do the dishes, or talk out a problem with Sue or Bill with whom I've had a misunderstanding.

According to William James, each of these self-assertive choices inscribes itself in the brain of the remote control user. By such handling of this device, the viewer is trained neither in concern for others nor in seeking the common good nor in exercising shared decision-making. Instead, such a TV viewer has increasingly molded himself or herself more and more into a self-assertive person who unilaterally expresses his or her power over the environment. He or she exemplifies Royce's first kind of attitude of will.

On the other hand, Royce's second criterion is fulfilled in those whose TV viewing renders them more isolated, lethargic, and merely receptive. Although studies show that the minds of TV viewers actively process, analyze, interpret, evaluate, and respond to the content presented, nevertheless, TV watching often renders viewers torpid, mesmerized, and almost trapped psychically, as in an addiction. Whether worn out by work or not, these viewers gaze at the moving images on the TV screen before them. They want to be entertained by being flooded with pleasant or exciting images. These provide the psychic world of someone else's images which retards the viewers from dealing appropriately with their own unique, personal reality. Moreover, these incoming images, as blended with the viewers' own activity, usually lead them away from their authentic center into their own fantasy-mixed worlds--not the actual world of inner and outer reality. Many such viewers use TV as their escape route into an artificial world.

This can also take place in a group way. Sociologists have identified a process of group-individualism which they call the "cocoon effect." In it people seek to escape from their real problems by looking for input that appeals to them, is acceptable to them, and makes them feel comfortable. As already mentioned, designers of TV shows are targeting increasingly specific audiences by using ever more specialized channels. In this way they fragment audiences into smaller and smaller interest groups that share little in common with the rest of society and are out of touch with the rest of the world beyond their small group.[38]

Thus whether individual or group, they become withdrawn from interactions with real people in the world. Psychically they often become solitary couch potatoes. And in the future the point may be reached in which people somehow plug computer chips directly into their brains, so that by themselves they can float along the flow of the World Wide Web or even enter the virtual reality of another person's experience, even while each is all wrapped up in one's own cocoon.[39] These cocooned individuals or groups exemplify Royce's second criterion of the isolated self withdrawn into itself.

Thirdly, however, there are many people who are committed in loving loyalty to serving the people and communities around them. Many of these very loving, giving, community-minded persons freely use TV, radios, cassettes and newspapers, but without addiction. They are engaged in serving others; they try to heal psychic wounds, to listen to others carefully, and maybe to teach literacy to immigrants, or to repair broken-down houses. They feel a realistic solidarity with their nearby neighbors and with the whole human family. Wonderful people like these are also careful about household tasks, family rituals, and practices. Among them are those thousands of volunteers, those "green sprouts" of an America on the mend. These are the people who put democracy into living practice and whose success stories are broadcast by *American News Service*. They are the lovingly dedicated people of Royce's third criterion.

In review of the last two chapters, we have discovered three kinds of influence and control upon, within, and by the media. Our media are dominated by the U.S. government and by big business--especially, through its advertising and public relations firms. Then the media's hierarchs dominate their own personnel. And the present chapter has shown that all these media-related forces influence and control the thought and value systems, as well as the behavior patterns, of ordinary Americans. Moreover, they similarly affect those billions of people living outside the United States, most of whom are caught in miserable economic conditions and are prey to the direct and indirect commercials of our exported media. So, while maintaining a primarily positive attitude towards TV and the other media, the need arises to focus in the next chapter on the know-how that gives the media its gigantic technological power.

QUESTIONS FOR DISCUSSION

1. Besides health, freedom, and democracy, what other positive values have TV and the other media helpfully impressed upon Americans?

2. Besides anti-communism, anti-Third-Worldism, and "America number one," what other stereotypes have TV and the other media fixed upon the minds and hearts of Americans?

3. How do the media inhibit Americans from understanding and appreciating people of other cultures?

4. In what ways is "America, number one"? How do the media promote this image? In what ways is America not "number one"?

5. How do you feel about the amount of violence on TV? Have programs with violence contributed to the violence in our society? What evidence grounds your response?

6. If certain kinds of programs should be prefaced with the caution, "This program may harm your psychic health," which programs would you nominate for this warning?

7. What is a gender stereotype? What are some attitudes toward sex fostered by certain programs and commercials? Do all these attitudes show disrespect for the embodied human person?

8. What is more important than the V-chip for protecting children from pornography, sex videos, and other erotica?

9. How can rash judgments and snap decisions in daily life be attributed to the media, especially to TV?

10. Americans are 5% of the world's population yet consume 40% of the world's resources. How have the media contributed to this unbalanced situation?

Chapter 9

ELECTRONIC TECHNOLOGY AND CORPORATE POWER

Thus far we have looked at various forms of control by and in the media. Government and industry restrain the media; media editors and owners control their staffs; media personnel dominate their audiences. In the present chapter, we want to investigate how these various forms of domination became established and how this domination currently tends to become still more widespread.

Dominant Firms

Even without direct ownership, one company can dominate other companies. By exercising its leadership role, it can determine much of the competitive environment within which all other firms in the industry must operate. Without the consent of the other firms, the actions of this industry leader will shape a particular market outcome. One can test whether dominance is present by asking such questions as:

1) Does the firm have some means of controlling the competitive environment?
2) Can the firm's policies determine the nature of the market association itself?

The behavior of dominant firms includes:

1) setting the going price,
2) deterring outside firms from entering the industry,

3) adopting a strategy for both its investments and their financing,
4) diversifying its product line.[1]

Half of the dominant firms are members of the Fortune 500, the largest corporations in the country. They are heavy investors in agribusiness, airlines, coal and oil, banking, insurance, defense contracts, automobile sales, rocket engineering, nuclear power, and nuclear weapons. Many have heavy foreign investments that are affected by American foreign policy.[2]

For example, Gulf+Western (now Paramount Communications, Inc.) along with its two dominate operations, Simon & Schuster and Paramount Pictures, owns more than 100 companies in such varied fields as auto parts, lingerie, farm supplies, sports items, tobacco, steel mills, dresses, home furnishings, carriers for nuclear waste, television, productions, etc.

Interlocking Directorates

Many Americans are often unaware of a more subtle instrument of control besides the status of being a dominate firm. Yet, a certain technique, the formation of interlocking directorates, concentrates American business in the hands of a few powerful people.

A firm's board of directors is legally responsible for the corporation. Until the 1960's, usually directors were chosen from management within the company. It was rightly believed that these directors would be most interested in the success of the firm. During the late 1960's and into the mid-1970's, there was a major shift in the way directorates were formed. This was a shift from "inside" to "outside" directors. These outsiders are employed principally by other firms but by being drawn onto a particular firm's board of directors, they create an interlock of this firm with the other company. If the members of various corporations sit on the board of a particular corporation, they both form interlocking directorates and extend their control. Direct interlocks take place within the same industry. For example, if a board member of General Motors sits on the board of Ford, this constitutes a direct interlock. On the other hand, if a member of the board at General Motors also sits on the board of Chase Manhattan Bank, an indirect interlock is formed.

Interlocks provide a substitute for ownership by providing ties between companies that allow for extended control by a few powerful people. Usually the directors are recruited by the firm's chief executive and can be counted on not to interfere with management decisions. Rarely do they attack proposals or even pose embarrassing questions at board meetings. They can benefit the firm by offering advice and counsel to management and they provide some stability, being empowered to replace the chief executive in event of death or incompetence.[3]

Before joining the Supreme Court, Louis Brandeis called interlocks an "endless chain." He wrote:

> This practice of interlocking directorates is the root of many evils. It offends laws human and divine . . . . It tends to disloyalty and violation of the fundamental law that no man can serve two masters . . . for it rejects the platform: 'A fair field and no favors.[4]

Compared to the comments of Brandeis, the report of the Federal Trade Commission seems weak. The report states:

> Thus, the inherent tendency of interlocking directorates between companies that have dealings with each other as buyers or sellers, or that have relations to each other as competitors, is to blunt the edge of rivalry between corporations, to seek out ways of compromising opposing interests, and to develop alliances where the interest of one of the corporations is jeopardized by third parties.[5]

In the United States, there is extensive interlocking, especially in corporate headquarters in urban centers such as New York, Chicago, and Los Angeles. The firms most likely to be involved in interlocking are large firms operating in capital intensive industries where the firm is controlled by diffuse interest groups.[6]

When Lydia Packard opened her small boutique in Chicago, she wanted to keep the business simple--make the necessary decisions herself, do her own buying and pricing, hiring just a few clerks to serve the customers. As the business began to grow, Lydia realized that it would be helpful to have a board of directors. She had been so busy building up her business that she had not kept abreast of changes in the formation of directorships. Little did she know when she recruited directors how naive she was about directorships.

Now she discovers that the directors on her board are also sitting on the boards of other companies in the fashion industry. Lydia finds that she is strongly advised on price setting and on her transactions with buyers and sellers. She discovers that these ties of direct interlock circumvent market constraints.[7] The competition that stimulated her business has all but disappeared and she feels she no longer really owns her own business.

Lydia's feelings of inadequate control of her business are not unique. They represent on a small scale what owners and executives of many larger industries must experience. As the political scientists Thomas Dye and L. Harmon Zeigler point out:

> More than 200,000 industrial corporations exist in the United States but the top one hundred control 52% of all assets in the nation. The five largest industrial corporations control 10% of the nations assets themselves. Concentration in utilities, transportation, and communications is even greater.[8]

Conglomerates

The controllers of most of America's communications systems are partners in industries such as agribusiness, airlines, coal and oil, timber, banking etc. A good number of these media corporations form the core or significant part of conglomerates--that is corporations which have business operations in a number of different business areas.[9]

As we found in Chapter 7, mergers, acquisitions and takeovers in the United States during the past twenty years have concentrated media ownership in the hands of fewer than fifty corporations. The presence of these media corporations on the *Fortune 500* list has tripled. These media conglomerates earn more than 50% of world-wide film profits, and control over 75% of all TV programs imported around the world.[10]

As media conglomerates have become larger, they have been integrated into the higher levels of American banking and industrial life. They become part of a conglomerate as subsidiaries or through interlocks within their boards of directors. Conglomerates are always on the lookout for ways to influence public opinion and government policies, both at home and abroad. What better way than to own most of the news media and extend their influence through interlocks?[11]

As early as 1979, a study by Peter Drier and Steven Weinberg found interlocked directorates in major newspaper chains. Gannett shared directors with Merrill Lynch (stockbrokers), Standard Oil of Ohio, (now part of British Petroleum), 20th-Century Fox, Merr-McGee (oil, gas, nuclear power, aerospace), McDonnell Douglas Aircraft, McGraw-Hill, Eastern Airlines, Phillips Petroleum, Kellogg Co., and New York Telephone Company.

Time, Inc. (even before it became Time Warner) had so many interlocks it almost represented a plenary board of directors of American business and finance, including Mobil Oil, AT&T, American Express, Firestone Tire and Rubber Co., Mellon National Corporation, Atlantic Richfield, Xerox, General Dynamics, and most of the major international banks.[12]

Through news reports, most Americans are aware that giant mergers are occurring. For example, General Electric merged with RCA in 1985, forming the fifth largest corporate merger in American history. Time, Inc. merged with Warner Communications. On two consecutive days in 1995, two mergers were announced that made former mergers seem almost insignificant. On July 31, Walt Disney Co. acquired Capital Cities-ABC Inc. for $19 billion. The next day Westinghouse Electric Corp. announced it had bought out CBS, the nation's last independent TV network for $5.4 billion.[13] Other mergers have taken place with hardly a ripple in public interest but these two mergers involving TV networks got immediate reaction from the public. Looking to their own interests, many people are asking, "How will these buy-outs affect TV programming? Will programs become more juvenile and trivial as Disney influences the ABC network?"

Even more than dominant firm status and interlocking directorates, mergers and diversification have concentrated the ownership and control of business in the United States. To avoid anti-trust suits, some corporations have so diversified that they own companies in many unrelated industries.

In 1990, the twenty-three corporations that dominated all major media were:

1. Bertelsmann, A.G. (Books)

2. Capital Cities/ ABC (Newspapers, broadcasting)

3. Cox Communications (Newspaper)

4. CBS (Broadcasting)

5. Buena Vista Films (Disney; motion pictures)

6. Dow Jones (Newspapers)

7. Gannett (*US Today*, and 87 other dailies)

8. General Electric (Television)

9. Paramount Communication (Books, motion pictures)

10. Harcourt Brace Jovanovich (Books)

11. Hearst (Newspapers, magazines)

12. Ingersoll (Newspapers)

13. International Thomson (Newspapers-120 dailies)

14. Knight Ridder (Newspapers--*Philadelphia-Inquirer*, *Miami-Herald* and 27 others)

15. Media News Group (Newspapers)

16. Newhouse (26 Newspapers, books, including Random House books and Conde Nast Magazines)

17. News Corporation, Ltd. (Murdoch; newspapers, magazines, motion pictures)

18. New York Times (Newspapers + 26 other media outlets)

19. Reader's Digest Association (Books)

20. Scripps-Howard (Newspapers)

21. Time Warner (Magazines, books, motion pictures)

22. Times Mirror (Newspapers)

23. Tribune Company (Magazines) [14]

By 1998, this number had grown even smaller.

Time Warner is the largest media corporation in the world. One gets a sense of the power of this conglomerate by considering several facts about Time Warner. It has more technical communication power than most governments. Its assets are greater than the combined gross domestic product of Bolivia, Jordan, Nicaragua, Albania, Liberia, and Mali. It is the largest magazine publisher in the United States, with a readership exceeding 120 million. It publishes, among others, *Time, Life, Fortune*, and *Sports Illustrated.* It is the second largest cable company in the world and one of the largest book publishers, including in its list Time-Life Books; Scott Foresman; Little, Brown; and Book-of-the-Month Club. It is also the world's largest video company in the world with subsidiaries in Australia, Asia, Europe, and Latin America.[15] One might well ask, in developed countries is it possible to avoid the influence of Time Warner?

Ted Turner and Rupert Murdoch, powerful competitors, run conglomerates that encompass not just newspapers or magazines or movies or broadcast TV or cable but all of the above and more.[16] In late 1995, Murdoch resorted to vapor news, an announcement intended to affect the market, confuse his rival, and cause customers to hold off buying the rivals goods or stock. Murdoch declared his intention to launch a round-the-clock news channel to rival Turner's cable news network. This was not a firm decision for the immediate future but a distraction from Murdoch's campaign to capture satellite-TV in Latin America. With MCI he is also working to challenge the on-line market of services like America Online.[17]

In the Fall of 1995, Kevin Kelly, the executive editor of *Wired*, predicted in an interview for the *New Perspectives Quarterly*, that Microsoft "is in a very precarious position" because of its bigness and its emphasis on software.[18] The alliance between Microsoft and NBC, however, seems to contradict Kelly's prediction. Bob Wright, president and CEO of NBC announced that it will be leaving America Online and Prodigy and joining Microsoft Network. The two companies will also collaborate on CD-ROMS, interactive TV, and other digital products.[19]

Cable TV

In the 1980's, cable TV systems were the most noticeable media development. It did not take long for cable ownership to follow the American pattern of concentration. As mentioned, ABC was acquired in 1985 by Capital Cities Communication Corporation, already a large corporation. The merger with ABC added 55 cable TV systems to the 12 TV stations, 11 AM, and 13 FM radio stations already owned by Capital Cities Communications.[20]

Another aspect of cable TV contributes to its ease of concentration. Because cable companies need right-of-way to run their wires, in most areas there is a cable monopoly--only one company is given permission by the local government to come into the area. Until 1986, local governments thought they could regulate rate hikes. The case in Vail, Colorado, however, demonstrated that power is in the hands of the cable monopoly rather than the local government. Telecommunications, Inc. asked for a rate increase in Vail and was denied. One weekend it shut down all programming and for 48 hours nothing appeared on the screen but the name and telephone number of the town manager in charge of the cable contract.[21]

Telidon in Canada, QUBE in Ohio, Prestel in Britain, Captains in Japan, and Antipope in France are primitive versions of what is emerging today--mass terminal systems, connected to large public databases and larger private information stores to supply the individual home with a huge variety of services.[22]

In 1986, cable companies gained further power when a law passed by Congress went into effect. This law prohibits cities from regulating cable rates. As a result, prices for service have soared 61% in those areas where there is no competition. Although 11,191 cable systems operated in June 1992, only 65 communities had more than one cable system. In these communities, bills are 22%-30% lower, consumers have more channel choices, and service is better.[23]

Bagdikian alerts us to the dangers the U.S. faces through the power of these huge conglomerates. He writes:

> The new global [media] giants have their own style of control wittingly or not. Given the dynamics of advertising and their drive for maximum audiences, they are also moving toward elimination of political and cultural diversity. The economic and political goals of these corporations are quite uniform, their output is remarkably similar, and their communications power is greater than that of most nations. They move internationally without even mild antitrust restraints of their home countries. Yet once their commercial empires have achieved monopoly-like power, they can control their profit levels and be free to pursue their own political and economic agendas with less restraint.[24]

The dangers Bagdikian mentions extend beyond the U.S. deep into Third World countries. The gap that now exists between industrialized nations with well-developed telephone systems and Third World countries with inadequate and poorly functioning telephone systems is widening at an alarming rate. Moreover, the rapid growth of computer and telecommunications systems is deepening the chasm between First and Third World countries.

As early as the 50's and 60's most nations were convinced by American corporate interests, as well as by the U.S. government and military, to accept the commercial model for TV broadcasting. These nations failed to question whether the model suited their country or not. They soon realized, however, that the cost of producing their own TV shows far exceeded the cost of American TV products which were shipped to foreign stations at cut-rates.[25]

Some undeveloped rural areas of the U.S. share the disadvantages of Third World countries. For example, parts of northwestern New Mexico are miles from the nearest power line and have no telephone service within walking distance. The Monastery of Christ in the Desert is located in such an area. Here the Brothers, unlike their neighbors and most people in the Third World, the Brothers have the creativity and the means to overcome the disadvantages of their location. They generate electricity by a dozen solar panels and with the aid of a cellular phone they are plugged into the Internet. Through the Internet and their own WEB page they are connected to the world beyond their monastery.[26] Most native Americans in northwest New Mexico, however, like their brothers and sisters in the Third World, lack any such connections.

As a sign of progress, Third World nations are persuaded to invest in hardware--TV's, VCR's, and computers. Little thought is given to the skilled technicians and infrastructure needed to support the use of this hardware nor its long range costs where leasing contracts are involved. Often less thought is given to selecting software appropriate for the language and culture of the country.

In an online question and answer session for *Time* magazine, Nicholas Negroponte, M.I.T media specialist, was asked: "Do you think the increased interaction between peoples of different countries over the Net will impact our various languages?" Negroponte answered:

> Yes, the Net will make local language trivial and a global language exist as well (probably English). Many Europeans and others see that like airport tower control (which is English except in Montreal!) But

> the stronger force is that the Net will ENHANCE local languages in parallel. It is just that the local may be smaller than a nation, closer to a tribe.[27]

As Anthony Smith looks toward the future, he writes:

> It [the deprived society] would be more helplessly locked inside the spiral of low-information generation than today. . . . The divisions between the information-rich and the information-poor --internationally and nationally--could become almost inexorable, far harder to overcome than the divisions founded upon economic exploitation.[28]

The Electro-Magnetic Spectrum

Beyond problems of hardware, software, telephone and cable systems, a major factor involved in the control and distribution of electronic information is the dependence of all information systems on electro-magnetic energy. Unlike mineral resources, many of which have already been depleted, electro-magnetic energy is the one non-depletable resource. This energy can be made to oscillate at different rates, that is, move in waves at different speeds. The resultant broad span covering the entire available range is called the electro-magnetic spectrum. Since electro-magnetic waves travel at the speed of light (186,000 miles per second), if such a wave completes its cycle in one second, it has the now-established basic frequency, of one hertz, the standard measure of wave frequency. A kilohertz measures a wave frequency of a thousand cycles per second; a gigahertz represents a wave frequency of a billion cycles per second.

Given these different wave frequencies, the question arises how can the various wave lengths be distributed fairly to the earth's actual and potential users. To carry out this fair distribution the International Telecommunications Union was established in 1965. As we shall see it has not been altogether successful in preventing imbalances from arising.

The power of information was very evident during the World Administrative Conference (WARC) in 1979 in Geneva. As a result of WARC, "it was accepted that developing countries should have more equitable access to frequencies and also that countries should be required to relinquish unused frequencies and clean up the Master Register.'"[29] The problems of distributing high frequencies could be settled fairly by expert planning. As expected, complex technical arguments arose at the Conference about how to organize the use of radio wave lengths in the electro-magnetic spectrum.

The U.S. delegation of one hundred people came to Geneva with a detailed plan. This plan, prepared by a thousand people, covered the entire spectrum from 1 hertz to 387 gigahertz. They also had available data concerning the positions of all the other delegations. All this information was on-line at computer stations and available at a moment's notice. In contrast, some Third World countries had only one or two delegates. For some delegates, it was their very first experience in telecommunications negotiations. These delegates were at a further disadvantage because they had decided to treat the problem as a justice issue rather than one of technical planning.[30] Faced with the U.S. power in negotiating, how could small countries with one or two delegates even begin to compete?

Another aspect of U.S. power in global communications was evident at the general meeting of WARC in 1979. France had a large range of fixed-band high frequencies which had been allotted to her but which were no longer in use because of technological advances. France made these frequencies available to nations that needed them. The United States had a much larger list of unused frequencies, registered for services that have been abandoned as new technology became available. These frequencies, however, were in the possession of private corporations who regarded them as part of their fixed capital. By hoarding unused frequencies, the United States deprived Third World countries of resources that could contribute to their growth.[31] Unlike other world resources, the spectrum allows of almost infinite possibilities. The more that electro-magnetic problems receive ever more sophisticated solutions, the more the less sophisticated users become dependent--at a cost--on the expertise of the highly developed countries.[32] Even a country such as Sweden depends on U.S. technology. There fire fighters depend on a computer data bank that keeps records of each building in a city. The records include the construction features, number of occupants, and any special characteristics that might be important to fight a fire. The data bank also indicates the fastest route from each building to the nearest fire station. Despite the 8,000 mile round trip for the transmittal of information, the contract was awarded to General Electric and the data bank was originally located in Cleveland, Ohio.[33]

As research confirms, technological growth does not necessarily ease the access of all peoples to national and international information.[34] The same trend governs information as it does economics: the rich get richer and the poor get poorer--a trend that holds true both of individuals and of entire nations.

During the past ten years, governments in both industrialized and developing countries have become more aware of the need to adopt national policies on communications and information issues. As the Egyptian communications scholar, Hamid Mowlana comments:

> Developing countries particularly view their dependence on foreign firms and transnational actors as evidence that the important basis for national decision-making is now located outside their national boundaries. Their national policies are being designed in order to preserve national sovereignty; to insure national security; to assure access to information held in data banks; to preserve cultural identity; to design appropriate educational, scientific, and artistic policies; and in some cases, to protect individual privacy.[35]

Such national policies express an ideal for developing countries that may be very difficult to approach. These countries can use the information available only if they have in place the needed infrastructure--e.g. roads and transportation, dependable electricity and computers, basic literacy, competent schools and colleges for technical training. This infrastructure has usually been in quite short supply in developing countries. Fortunately, costs of microprocessors and telecommunications systems have recently decreased significantly--particularly the cost of access to information via satellites. If they plan carefully, then, Third World countries need not go through all the stages of technological development experienced by industrialized nations. Yet these Third World societies need national policies that prevent industrialized nations from discarding cheap and outdated equipment on them.[36]

Informatics

A special word, "informatics," has been coined for the organization of equipment for the new information services. Informatics includes the marketing, investment, research, and manufacturing which result in a society being provided with the means to collect and distribute information.[37] Increasingly, information is being viewed as a strategic resource that enters into calculations of national strength in both military and industrial espionage. Nations battle to use modern information technology most advantageously.[38]

Data can no longer be isolated into categories required by government, or industry, or banking, or newspapers and magazines.[39] As Anthony Smith puts it:

> Modern systems for ferrying data carry the medical, criminal, employment and credit records of individuals; they also carry the data necessary for modern industries, including credit data of private and public companies; they carry the information required by government to plan development. The strands are inextricable, and the companies which control the manufacture of the equipment (which are very few in number) hold the key. There is no longer such a thing as national autonomy without control over data flow, and informatics. However much a nation--or all nations--desire to reverse the inequalities in the flow of international news, their intentions are doomed without autonomy in the field of informatics.[40]

Herbert Schiller, a communications researcher at the University of California, criticizes American communication practice. It not only exports products and services in this field but, more importantly, it exports a system. The receiving nation obtains American values and attitudes that are often not in that nation's best interests.[41]

Wilson J. Dizard details this problem by saying:

> The most controversial area of U.S. information sales abroad involves film, television, publications and news agencies. Here the American information age is made visible to literally billions of people abroad in ways that are brash, colorful and immensely popular. It will be generations before the impact of contemporary U.S. culture on the rest of the world can be measured. Will it be judged as the raucous noise of a flash-in-the-pan society, as critics claim, or will it be seen as the transmitter of ideas and values that created a new democratic world order?[42]

The company holding the key to informatics is IBM. As early as 1980, it was responsible for 70% of all computer installations in the entire world. Already in 1978, its revenue from the rental of computers was $10 billion and the profits from this sector accounted for half of its entire profits. In a rather vain attempt to maintain some control, the U.S. government has refused permission to IBM to sell computer services such as data processing. This restriction, however, does not apply to overseas trade. In this way, IBM gains another $1 billion a year through the sale of computer services abroad. It employs 130,000 workers in plants outside the United States. In nearly all of the developed countries IBM has at least 50% of the total communications market. Even in Japan, which is well on the way to self-sufficiency in computer hardware, IBM still has over a quarter of the market. It is the power of IBM which helps to explain the extraordinary predominance of the United States in the information field.[43]

IBM, however, faces many problems in maintaining its international position. France was the first to decide that IBM's near monopoly had to be ended. This resulted in a cycle of mergers, licensing deals, and other arrangements, all designed to create a workable domestic alternative to IBM. IBM was portrayed as the villain, the strongest factor in American domination of French computer resources, and the data banks on which French industry depended.[44]

Soon, the European Economic Community followed France, motivated by the fact that the EEC produced only 15% of the computers and could account for only 26% of the consumption. More recently Japan has entered aggressively into the international high-technology market. Their competition will be increasingly felt in the next few years.[45]

One very striking area of development is remote sensing by photo-satellites. Under the Freedom of Information Act, American corporations are able to use the knowledge from these photo-satellites for their own benefit. For example, NASA's Landsat set of satellites is able to collect information about the resources of developing countries without communicating it to these countries. Any Third World country can no longer retain strict secrecy about its mineral deposits, its harvests, its distribution of land, or any of its material resources. It might not know of the presence of certain valuable mineral deposits within its borders. United States' mining companies may hurry to take options on tracts of land before the local government has awakened to its mineral riches.[46]

The Geo-Stationary Orbit

The electro-magnetic spectrum interconnects with another world resource: that precise orbit for "parking" satellites 22,000 miles from the earth. If properly parked, three satellites can simultaneously send signals to the entire world. In the few years since space communication became possible, so many satellites have been sent up to park in that particular orbit around the earth that it is in danger of becoming cluttered. Although this geo-stationary orbit embraces all nations in common, currently only a handful of nations possess the expertise and the wealth to launch their own information satellites into this orbit. This gives tremendous power to these more developed countries, while reducing Third World countries largely to passivity. Moreover, American satellites and those of client nations are demanding ever more parking spaces within this very convenient orbit. In this way, the Western nations are upping the ante for a Western domination of the globe through information. Nevertheless, the geo-stationary orbit and the electro-magnetic spectrum remain the common property of all earth's inhabitants, like the oceans. Unlike oil, coal, or gas, these global resources can neither be exhausted nor increased by human effort. They need to be utilized, without favoritism, for the enhancement of all nations of the human family.[47]

In this chapter, then, we have surveyed some of the instruments producing the media conglomerates' ever-widening network of domination. We measured the effects of mergers, diversification in holdings, and interlocking directorates of the 23 media giants. We examined the widening spread of cable TV in competition with private and public TV networks, vast anonymous data banks, and America's narrow-minded possessiveness of both the electro-magnetic spectrum and the geo-stationary orbit for satellites. In the next chapter we will consider how the present situation might develop in the future.

QUESTIONS FOR DISCUSSION

1. Discuss the ways companies can gain control of other companies. Do you think this trend is good for ordinary citizens?

2. Do you agree with Louis Brandeis that "the practice of interlocking directorates is the root of many evils"? What specific evils do you perceive?

3. How has cable TV contributed to media power?

4. How does U.S. power over global communications affect Third World countries?

5. What are some of the limitations of Third World countries that prevent them from sharing in the information glut?

6. What part does IBM play in U.S. control of information?

Chapter 10

THE MEDIA'S IMPACT ON THE FUTURE

The previous chapters of this book revealed the media's current structures, power and tendencies, their present potentials, dangers and promises. Surprising electronic discoveries and novel applications to the media provide the roadbed for the new information highway.[1] Yet how will all these changes influence the future? How will this highway affect global mass culture and its civilizations? What impact will it have upon upcoming politico-economic changes around the world?[2] Specifically for us Americans, how will this inrushing cyber-age affect our children's lives and American culture in the year 2020--less than a quarter century from now? These questions supply the thread unifying the present chapter. They also shift the authors into predictors of the future, although they know no one can accurately predict the surprising future.

As the 20th century started, the change from horse-drawn buggy to automobile changed people's behavior and their culture. It not only spurred entrepreneurial global grabs for supplies of oil and steel, but nearer to home, profoundly changed urban, business and family life. Yet today these once astounding changes seem small indeed compared to the changes awaiting our children and grandchildren. A change in transportation touches the human psyche less intimately than a change in communication. So, because the electronic revolution of the cyber-age brings changes into our ways of communicating, its effects will be most profound.

For instance, as VCRs enter more and more homes around the globe, not only will the popularity of American movies on cassette increase dramatically,[3] but Hollywood's attractively packaged values will be proposed to billions as to what is smart and right. In 1996 the households of 15 million North Americans had online access to the Internet. In just four years, by the year 2000, the households of 38 million North Americans will have online access to the Internet--most using it for browsing and entertainment.[4] With so many people seeking the leisure of being amused and entertained, will America's work ethic be rendered anemic or even starved to death? Whether so or not, some in-depth change must be causing such a widespread transformation of households as this..

The basis of our communication is undergoing a profound shift from an analog system (based on electronic wave lengths) to a digital system (based on electron bits). At bottom it moves from resting on atoms to relying on infinitesimal electrons. That is the way Nobel Laureate Arno Penzias described it. In late 1996, with FCC mediation, the TV and computer industries agreed on the technical standards required to make digital TV the basis for home entertainment by 1998. The result will be wider screens, high density pictures, CD quality sound, clearer reception, and much more information.[5] This shift from an analog to a digital system causes a fountain of changes to spray into sight. People will move more and more from the fixity of print to flow along on the streams of orality. In cable TV, the 500 or more highly specialized channels will appeal to individual interests while holographic TV, holovision, may be in living rooms in a decade or two.[6] The technology of the digitally-based World Wide Web already fosters commerce by easily identifying prospective buyers of stocks, autos, houses, books, and small objects through a process that essentially resembles mail-order retailing. For instance, General Electric predicts that its 1996 online sales will reach almost $ 1 billion. So many business transactions might lead one to ask whether the Web will become more commercial than informational. Yet this overlooks the fact that currently the Web is used far more extensively for browsing and entertainment.[7] The shift to digital will also show itself in the linking of TV with computers so that Internet TV sets and Internet telephones will render traditional TV sets largely obsolete. By the year 2000, then, people will be amusing themselves even more by playing Web-based video games.[8] And in the field of radio, digital audio broadcasting (DAB) will result in far higher quality radio reception. Its advent will remove the differences between the two analog-based forms of radio broadcasting--AM and FM--and result in a restructuring of the radio industry.[9] Then, too, movie-making will be done by shooting scenes digitally onto film--thus eliminating the developing and shipping of films.

Movie-viewing, beyond using video cassettes on a pay-per-view basis, will shift onto a movies-on-demand basis in which the viewer has a far greater selection of films and can choose when the movie is to start.[10]

These instances of the results of the profound shift from an analog to a digitally based communication system suggest how deeply the methods of business and government will be transformed. Even more concretely, as mass culture is recast by this new mode of digital communication, so too will the various civilizations of the globe be refigured, insofar as the latter rest on mass culture.

Gazing into the future, then, predictors have the task of discerning the distinctive trends of this new future shaped by present and upcoming discoveries in electronics and their innovative applications. Empowered by these emerging electronic technologies and the adventure-filled information highway, the digitally transformed media will have massive impacts first on the business world, then on government, and finally on mass culture and the civilizations dependent on it. These three areas of media effects suggest the order of the present chapter.

The Media's Impact On the Business World

Business life is fundamental to human existence. First, then, how do the digitally transformed media affect the business world? Here business can be examined from the viewpoints of economics, finance, commerce, and warfare. The last approach is not surprising, since in the long run warfare has usually stimulated free enterprise. The cyber-age may well have its widest and most pervasive impacts in the realms of economics, finance, and commerce .

Impact on economics, finance, and commerce

At a supermarket no sale escapes the check-out counter's electronic eye which registers goods sold and payments made by consumers' credit-cards. Supplied with this data, computers provide an evermore exact measurement of the ebb and flow of supply and demand. Thus, more cleverly than local bookkeepers, computers better steer the huge vessel of world commerce. On this electronic basis, orders or cancellations are flashed to suppliers, and truckers are directed to certain specified loading docks, with just certain products, to arrive at just certain times. The same electronic network controls salary computations, price changes and labeling devices.

Strangely, the cyber-age may undermine belief in the adage, "The bigger, the better." For the importance of economies of ever larger scale seems to diminish as more specialized markets are sedulously cultivated. Thus, media experts foresee that big networks will largely be replaced by smaller, more customized offerings.[11] Moreover, the need for traditional office jobs with their paperwork of endless typing, sheet-shuffling and filing, is becoming outmoded. Meanwhile, the trend toward, and demand for home-site jobs, carried out by computer, modem, and fax, is sharply increasing.[12] Paradoxically, then, the cyber-age may counter the industrial revolution's tendency to require workers' presence in factories and offices, a practice which sharply reduced family time together. By its promotion of cottage industries, the cyber-age may allow family members to be at home more, to have more opportunities for family interaction, and thus may even contribute to family bonding and union. Supply-side economist George Gilder even predicts in his book, *Life After Television*, that networked personal computers will transform capitalism into "a healing force in the present crisis of home and family, culture and community."[13] Yet, as Joseph Dominick predicts, the costs of genuinely high quality TV programs may insert a dangerous wedge between the rich and the poor both in the U.S. and in other developed countries:

> In the long run, we run the risk of creating segments of society that are "communication rich" and those that are "communication poor." Children who grow up in households that can afford computers, educational tapes and videodiscs, cable TV, and pricey magazines and books might have an insurmountable head start in school over those children not so privileged, further increasing the fragmentation of society. Would taxpayers be willing to subsidize "information stamps," much like they do food stamps, to aid the underprivileged?[14]

Because of a media-transformed economy, in the near future people will use touch, as well as their eyes and ears, to operate their personal computers both at home and in business offices. Already they use touch to open some doors, elevators and hand-drying machines. But this time the touch will be by human voice and its waves. Here is what Negroponte forecasts:

> In the same way that we reach out to touch someone we will find ourselves using voice to project our desires to machines. Some people will behave like drill sergeants toward their computers,

> and others will be the voice of reason. Speech and delegation are tightly coupled. Will you be issuing orders to the Seven Dwarfs?
>
> Possibly. The idea that twenty years from now you will be talking to a group of eight-inch-high holographic assistants walking across your desk is not farfetched. What is certain is that voice will be your primary channel of communication between you and your interface agents.[15]

What will be the impact of voice initiation on its users and hearers? Some people will tend to use voice initiation out of self-aggressiveness--like drill sergeants, as Negroponte puts it. With a "don't bother me" attitude, others will speak from an escapism that seeks comfort. Still others will speak compassionately and cooperatively out of a realistic commitment to promote the common good of all, as far as possible. So, voice initiation will make manifest who are the world's unilateral attackers, who its cowardly comfort-seekers, and who the dedicated struggling servants of the human family.

Meanwhile, in the financial world, thanks to computers, mega-millions of dollars (or yens, marks, and pounds) are being daily exchanged across international borders, as investments are made, withdrawn, or redirected. Since everyone in the world of finance is so busy making more money, few raise some annoying questions. Will anonymous cyber money create a black market money economy? Will electronic investors weigh the impact of their moves on the global environment: increased pollution, deforestation, greater global warming? Will electronically super-sophisticated insurance companies peer further into people's private lives? Should some tariffs accompany the international transfers of funds? After all, imports from foreign countries are charged with tariffs or custom duties. Similarly, should a U.N. tariff or custom duty be levied on major money transfers as a contribution to the development of the economically least advanced nations? Such nations lack both the developed nations' level of electronic communications--including access to global satellites--and even a minuscule percentage of those millions of dollars traded every day. If these impoverished nations were supported by such a tariff, their too heavy burdens of foreign debt would be lightened considerably. At present, however, the electronic cyber-highway is only widening the gap between our globe's affluent North and destitute South. This diversification grows even sharper than that affecting the "communication rich" and "communication poor" in the U.S., as just described by Professor Dominick.

Impact on warfare.

Many futurists argue that global war is no long viable because it is a lose-lose situation. Rational as this is, the globe has had its irrational leaders--its Hitlers and Stalins--and still has agents irrational enough to terrorize millions to destabilize the world's present economic system or simply to get power. So the Department of Defense at the Pentagon cannot plan as if everyone were rational.

Militarily, then, the electronic revolution has launched us into the "Age of Cyberwar."[16] World War I introduced tanks, trench warfare and poison gas. World War II brought electronic decoders and nuclear bombs. In these wars, the total number of those killed leaped awesomely from 8 to 60 millions. For the next war, however, strategists are relying on cyber-age technology. They will jam enemy radio transmitters with misleading signals. They will insert viruses into computer-based telephone systems. They will plant electronic impulse bombs into the communication systems that govern enemy transportation and finance centers, thus creating unprecedented panic and terror. They will do all this well before the enemy fires his first shot tipped with atomic, or biological, or chemical poisons. Confronted with this vision of a new Armageddon, one wonders whether the number of those slain in the coming cyber war will indeed fulfill that infamous 666 in millions.[17]

In this new style of warfare, each infantryman will be electronically equipped with the super-sensitive gear he will need. He will wear night-vision sensors, a video-panel, voice- activator, and a wireless governed gun-aiming device. It will cost tens of thousands of dollars to equip each infantry person with such armor. Leaving unmentioned the increased profit accruing from sales of electronic equipment, one sees that the competition to supply the super-sophisticated electronic gear needed for such warfare will expectably surpass by far the current competition to sell new tanks, armored carriers and planes.

For the United States, however, the stakes and risks of cyber war are all the higher since it is the most electronically dependent country in the world. The United States is especially vulnerable because all its activities--those on Wall Street, at airports, research centers, and those found in homes, schools and farms--in short, all the functions of its highly refined infra-structure, rely heavily upon electronic equipment. Yet even now, despite protective devices, some ingenious high-school hackers can, with skill and chance, succeed in entering corporate electronic systems.

How much more so will sharply-trained electronic terrorists plan assaults far more devastating than those visited upon New York's World Trade Center or Oklahoma City's Federal Building? Here even the optimists paint a bleak picture because the United States is far more vulnerable to electronic sabotage than are less developed countries. In this sense, America's great strength in electronics may flip-flop into its greatest weakness, unless effective electronic gate-watchers are on duty even more carefully than the U.S. teams of SAC bombers have continuously been on guard against ICBM's.

The Media's Impact on Government Functions

Next, what impact will cyber-age and its accelerating rate of yet newer electronic applications have upon those governmental control functions found in law-making and political movements? How will the media revolution affect the making and enforcing of those regulations needed to establish order in the new open spaces? What results will the information highway work upon the electromagnetic spectrum and outer space, the ocean beds and such vast ecologically significant stretches as the wilderness of Siberia or the Amazon rain forests?

Legally, the electronic media have created and are creating a field-day for lawyers and a quagmire for international lawyers and members of Congress. Even after President Clinton signed the new Communications Act of 1996, which Senator Robert Dole called "corporate welfare," members of Congress are beset with lobbyists pushing various amendments to this media law. Battles rage between promoters of cable and network broadcasting, between libertarians and family-value proponents. As a safeguard needed by parents for the right rearing of their children, anti-violence and anti-porno groups insist that the V-chip be available for all TV sets and mandatory in new models. Against them, advocates for free speech and the ACLU protest against needless restrictions. Many legal experts see the need to revise the recently passed copyright law so that those billions of slippery electronic bits can be caught within the legal fingers of intellectual property owners.[18] Meanwhile, in many less legalized countries of the world, rampant piracy of CD Roms and other recordings has been going on.[19] In December, 1996, however, negotiators from some 160 countries met in Geneva and agreed upon the most sweeping extension of international copyright law in 25 years. Although other countries and the U.S. Senate still need to ratify these two new treaties, they counter cyberspace piracy and protect art, software, and music on the Web.[20] As a needed counterbalance to protect user freedom on the Internet, these final treaties also allow WWW browsers to make temporary copies of copyright material without violating copyright laws.[21]

Although the Internet's electronic bits amount to less than one percent of all the bits speeding through global networks today, the Internet is mushrooming so fast that experts predict one billion people will be using it by the year 2000.[22] This already occasions questions about how most fittingly and wisely to control mass media companies. Since the Internet is an interest of the entire global community, what are the public responsibilities and rights of media companies in the electro-magnetic spectrum? With the two international treaties of 1996 still far from being ratified globally, how is intellectual and artistic property to be rightfully respected in the electronic age? At both national and international levels, the increasingly pervasive effects of the Internet and other recent cyber-age phenomena will largely shape future legislation on the media. The recent U.S. regulation striking down a required duopoly of media in the same local region aims to evoke more competition. At the national level, some questions arise for the Supreme Court which years ago in an anti-trust move broke up A.T.& T. as a monolithic monopoly. Do the mergers of Time Warner with the Turner complex, of GE with NBC, of Disney/Cap Cities with ABC, and of Westinghouse with CBS constitute an oligopoly contrary to the intent of the anti-trust law? Will the Court soon move against these and other vertically integrated giants? The Court may find it wiser to promote, not further anti-trust decisions against these media behemoths, but regulations that effectively provide for public access and such diversity of programs as will ensure that anti-competitive forces do not arise from these alliances.[23]

The Supreme Court has recently boosted these giant media companies' financial interest in syndication--that is, in their "finsyn." Accordingly, new FCC regulations will release these media giants from their previous legal bondage to gather income from only the first run of any programed series. Now it will also allow them financial returns from all syndication of these series--that is, financial rights not just to the first time broadcast of program series, but to all their subsequent usages, whether in reruns, audio cassettes, foreign translations, and in all other commercial transformations of the first run of their programmed series. This will enhance income for the giants holding "finsyn" rights. It will put tighter control of distribution of movies, plays, and video books into fewer hands. It will further drain the media-colonies of the Third World which are already overburdened with debt-service on capital they need for bare sustenance and development.

Those without legal expertise wait to discover what national and international law can and will do to media mega-giants like CNN, Time-Warner, or Murdoch's Fox Communications. They likewise wait to see how much legal muscle national and international lawyers and courts can exercise to protect vast spaces like the recently opened-up Siberian wilderness or the ecologically endangered Amazon rain forests. In Siberia, for example, electronically guided extraction corporations like Exxon, Shell, Texaco and various timber and mineral companies are on the verge of committing the greatest rape of mother earth in history.[24] Out of respect for our only earth and for the good of future generations, the directors of these companies need to be reined in promptly and stringently by effectively enforced international law.

If attention shifts to outer space, Star Trek's continuing popularity is instructive. Such widespread feasting on science-fiction travels to galaxies beyond the solar system seems to create a self-fulfilling prophecy. Of course, the project of building stations in space and of astronomically exploring new galaxies billions of light years away will be computer-controlled and electronically-guided. These new bases in space excite people's curiosity and invite them to possibilities undreamed of in the days before 1960. Already in the '60's, earth-viewers were deeply touched by the first vision of their global village when early astronauts first transmitted home the TV picture of our only earth floating against the background of black space. How much more have countless TV space-programs pictured humans plowing ahead through space and hurtling safely between stars as they speed on toward unknown galaxies. Today even some humble computer-users, viewing their machines in the at rest mode, behold this vision of traveling through space. Surely such media-created scenes powerfully nurture the dream of space exploration.

Accordingly, by considering the most reliable and sophisticated electronic data gathered from outer space, specialized lawyers need to draft new international law and policies about space. They need to insure that outer space be cleaned up regularly and not become cluttered with debris. Their law must enable all nations to participate equitably in space exploration. Their law must distribute fairly to all nations the benefits deriving from all explorations of space and from all experimentations carried on in space stations. For as a resource common to all earth's inhabitants, outer space, like the ocean beds, is primarily common public property. Hence, any claims for privatization must at least be subordinated to the prior claims of this common good of the human family. The international public interest must be the predominant value which the new law insists upon while it downgrades global competitiveness and economic efficiency into service of this overriding value. Privateers' greedy exploitative grabs must be forcibly forestalled because they fail to heed the rightful priority claims of the U.N. as a whole and of future generations.

Media's Impact on Culture and Civilizations

Culture is founded on language and education and embraces the arts and sciences. So, the first consideration will be on the fundamental impact made by the new electronic media upon language and education, then upon the arts and sciences in general, and upon psychology and medicine in particular. This survey of the media's impact on culture and civilizations can then close by examining the new media's foreseeable effects upon the fruits of the arts and sciences: upon philosophy and morality, upon religion and life in religious groups.

Impact upon language

In a famous image, William James compared the human stream of consciousness to a bird's successive flights and perchings.[25] Following this rhythmic pattern, both oral and written language combine continuity of flow with certain featured constants for the resting places of thought. Yet when language is transformed electronically into a directional vector and thousands of bits, it means that language has reached a new center of gravity in its evolution. Previously, language had rested on the tongue's basic sounds which were carried symbolically by characters such as are found in the alphabet or pictograms. Now language's center of gravity rests increasingly on electronic impulses--symbolically carried by 1's and 0's united by a vector indicating some definite intent. Without jettisoning sound-symbols, numbers and logical relations--the old foundations of tribal or national languages--the new electronic language creates extremely swift energy symbols beneath all of these. It produces billions of electronic bits and rapidly combines these into sets of electronic impulses, called bytes. By intending to send or omit these as certain signs, it creates the new foundation for a language that any embodied mind can understand, whether here on earth or on some as yet undiscovered planet. Such electronic language tends to realize Teilhard de Chardin's prediction that this globe will become encompassed by a new kind of atmosphere--one of meaningful communication. He called it the "noosphere" and saw it could connect earthlings with minded beings anywhere in other galaxies.

English-speaking electronic engineers largely pioneered this shift in language's center of gravity to the bit level. As a result, the new media spread a dominant coating of English-usage throughout the global village. By using English, the electronic media broadcast the cultural strengths and weaknesses of Anglo-American civilization. Electronics, the technological carrier of this English language, is not to be faulted for the kind of spirit now endemic in the English language. Rooted in the excessive individualism of Hobbes and Locke, the spirit of the English language propagates an individualistic self-enterprise. This spirit wants to seize advantages unilaterally and treasures economic profit-making more than caring for any and all weaker, as well as future, members of the human family. It needs correction by the spirits of those languages that are based more upon family and other communities. One result of this is that despite computer manuals written in Japanese, French, Russian, Chinese or Hindi, people from these civilizations wishing to surf the Internet succeed largely to the extent that they can deal with English. Even more second-classed are peoples of the so-called Third World, especially in the impoverished areas of Africa and Latin America. The media-sponsored growing predominance of English worldwide may effect some global sense of human unity. The cost, however, may be a global deafening to the rich variety of different cultural sounds and values in other languages.

As hinted, this dominance of English-language in the media threatens the loss of the distinctive gifts offered by the French, Hispanic, Chinese, Russian, and other civilizations. For instance, the French insistence on intuition healthily offsets rationalism, the Hispanic emphasis on family holds individualism in check, Chinese indebtedness to ancestors counters now-generation Americans, and the Russian Orthodox sense of "from above" (as in their feel for *sobornost*) inserts a caution to over-confident secularists. These gifts are needed to create an emerging pluralistically enriched, global culture. Moreover, besides obscuring or even rendering obsolete many cultural treasures hidden in these different languages, the media-dominance of the English language often induces silence in members of non-English civilizations.

Impact on education

Grade schoolers in electronically developed countries no longer look up to alarm clocks on teachers' desks to find the time but glance at super-accurate, multi-functional, digital watches on their wrists. So they get one more jump ahead of--and separated from--Bolivian and Sudanese kids sitting in school on dirt floors. Many American grade schoolers raise their eyebrows at the old-fashioned way of adding and "doing figures" by hand when they can simply punch their sophisticated battery-run calculators to get their answers. So they become less equipped to count scores in non-electronic games or figure by themselves what amount of change is due after a purchase. Meanwhile, these youngsters' TV shows and record-albums stimulate their psyches far more frequently and pervasively than their teachers can do, whether in kindergarten or graduate school. So these students are invited to value TV and rock stars more highly than their teachers.

In line with this cyber-age revolution and with present requirements at some universities, the rule soon will be: "No admittance to high school and especially to college without at least computer competence in word processing." This rule will be needed to fit the younger generation into the cyber-age. Already thousands of grade schoolers are doing their lessons by computers. In sum, if this is the school situation in the '90's, what electronic devices can we reasonably expect in place for the education of students in the year 2020?

Libraries are already being transformed by online cataloguing, CD-rom reference sources, and digitally available newspapers and periodicals. Books may become printed on an on-demand basis from the publisher directly to any bookstore with downloading and computer-printing facilities. As a person uses a Sony "Walkman" to listen to music and novels, it appears that soon one will use a Sony "Bookman"--fed by a CD card that contains a whole book and displays one page at a time for the reader--hopefully equipped with accompanying sound should one prefer to listen.[26]

Given the way individual students are already working at their individual keyboards and video-games, the educational impact of such devices needs careful assessing. Without a skilled supervisor, it seems ambivalent. When at their keyboards, these students exercise choice, analyze situations, make judgments, and exercise creativity in their responses. At their keyboards they are playing, and thus preparing for life's more challenging circumstances. Yet, if the activities of each student are not inserted into the dynamics of a social group, such keyboard workers may foster habits of having one's own way, of an exclusivistic self-assertion, and an ever-greater individualism. In such an individualistic keyboard setting, these students do not have to adjust to the needs of real persons around them, except insofar as they meet with the designers of the unfeeling programs they are using or link up with electronic pals in cyberspace.

Moreover, many of these students' homes have three or four television sets so each family member can watch the program he or she wants. So, without being bothered to adjust to the wishes of other family viewers, each prefers one's own way of choosing and thus individualistic habits are deepened. Meanwhile, media experts talk positively about interactive TV enriched by much group participation and studies show that interactive media actually promote meaningful exchanges. Yet TV directors, conscious of public responsibilities, firmly insist on their controls against too much audience initiative, too open an access, and against interpersonal exchanges that become too confrontational.

In the electronic age, students in many American schools become swamped with a series of rapidly flowing impressions. To them the rule seems, "Get patterns; give immediate reactions!" rather than "View the thing in its historical context!" or "Tally up the elements of the situation to form a reasonable solution!" The two latter rules take time to carry out, while many TV programs and commercials call for almost instant response, approaching the speed of light itself. Admittedly, studies are clear that the media are neither the sole, nor even the primary cause of this lack of poise and reflective skills. Yet a complex of contextual factors have led in most American schools to a general reduction in reading skills and then in reasoning skills. Consequently, to preserve whatever wisdom lies in our American, and Western, and global past, a new elite of literati must be created using both ancient languages and video data. These will be entrusted to preserve personally the best of the past so that, to a mass audience increasingly forgetful of the past, they can witness the wisest insights of the past. They also need to make fitting suggestions for suitably adapting these insights to the new opportunities of the in-rushing future. Nevertheless, the egalitarian resistance toward creatively educating such an elite group is so great that in the foreseeable future there seems little chance of creating and employing such a group of literati.

Impact on the arts

Growing from language and education are the arts and sciences. So one wonders what impact the cyber-age will have on artists and their arts. What openings of the artist's subconscious energies will the new media unleash? What communal musical creations will be created by having data bases so easily accessible? The answers to such questions as these lie at least partly veiled because it is harder to gage how the new information highway will affect the free creativity of the artist. Hollywood will, of course, continually enhance its ability to render special effects. Also, photography will become more an art of greater timing and accuracy. It will also blur the line between photographs of reality and those artificially composed by some artist's skill--to leave in confusion consumers, jurors, and audiences generally. Beyond these novel techniques, however, it is not easy to predict accurately how the electronic age will affect artists and their works.

Impact on the sciences

Cyber-age enthusiasts dream of creating hitherto unheard of marvels and generating a thousand Einsteins. They aim to do this by harnessing that "fecundity of aggregation" of great minds, which a leading American thinker, Charles S. Peirce, identified. This attractive dream cannot be realized if scientists expect such results to arise primarily from the nearly infinite capacity of computers to code and store information--even computers far more powerful than current models. Computers' monumental capacities have to be ensouled. For this, the human power of serendipity is also needed. Needed, too, are the gifts of recognizing ambiguity and exploiting it, of creatively imagining untried paths and mentally exploring them. Needed, furthermore, are the talents to go fishing in the waters of vagueness for new possibilities and to see the new opportunities latent in different ideas, cultures, and competing minds. Once such initiatives are launched, then the fruit of these gifts can animate the promising infrastructure of the cybernetic age so that the dream of a thousand Einsteins can be realized. Then a community of morally sound and genius-level minds can converge in approaching solutions to our increasingly challenging social problems. Then, by employing the talents of many countries, a new universally valid creation can emerge that will nurture the coming of a richly diverse, yet basically unified, world community.[27]

Future scientists will vastly expand the power of the Internet or its successors so that standard procedure provide practically instantaneous consultation with fellow researchers in distant nations. Surely part of our electronic age will include further investigations into sub-atomic particles and anti-matter via improved super-collider and super-conductor instruments. Through the use of space-station telescopes, it will also include research into black holes and the birth of novas. In physics, the electronic revolution will lead us closer to solving the enigma underlying Isaac Newton's query about light--whether it is something beneath our interpretations of it as wave, particle, and field?

In applied physics, and despite formidable barriers in the developing but debt-ridden countries, the U.N. goal of a telephone within an hour's walk of every village in the world will soon be realized. Moreover, improved electronic microscopes will reach new levels of magnifying information. They will use their micro beams to situate far more than one hundred billion bits of information on ultra fine silver haloid film for an ever more accurate storage of data. In biology, electronic microscopes will probe deeper into the structure of genes to discover new possibilities They may even unlock plant chlorophyll's mysterious secret of synthesizing life from sunlight.

Social historians like Daniel Bell rightly warn people not to attempt a grand design of the future because of the elements of chance, risk and luck in scientific advance. Nevertheless, given the ever-accelerating rate of change and the increasing convergence of discoveries, the prediction can be made that present trends will pale in contrast to what the future will bring. And since science and art are often combined in the practice of psychology and medicine, the next need is to examine how these two disciplines will be affected by the electronic media.

Impact on psychology

The power of the mass media upon psychology is enormous, for weal and woe. TV shows like *Cosby* or *Touched by an Angel* and a movie like *Apollo 13* have focused millions of psyches on healthy family dynamics or uplifted them with the courageous actions portrayed. Yet Americans also recognize how feeding on the mindless fare found in most "soaps," crude Westerners, and cheap movie reruns tends to vulgarize taste. In the mid-90's, a strong and widespread complaint about violence on TV burst out against Warner Brothers specifically yet also against the TV industry as a whole. This complaint demonstrated how many Americans felt that their psychic nerves had been rubbed raw by what they perceived as a deluge of sex-and- violence. Parents found it flooding into their living rooms and poisoning the sensitive minds of their children even more than their own more calloused minds. As mentioned earlier, *Big World, Small Screen*, the American Psychological Association's five-year study of television's role in American life, documented this deluge of violence on TV. Also, children's' TV shows in the early 90's portrayed violence ten times more frequently per hour on Saturday morning programs for children than adults received during regular adult prime time evening programs. Thus, youngsters became ever more calloused both to a passive watching of violence being inflicted and to an unchecked active use of violence as the quicker way to solve their problems than dialogue.

Capitalizing on people's interest in the unusual and bizarre, some movies and TV shows horrify their viewers with real or fictional shockers. These blend power-crazy violence with crude sex, and perhaps with adventurously successful criminals, plus or minus ghosts or vampires. Caught by native curiosity, young minds, as well as minds not so young, drink in this potent brew. They measure neither how it dulls their conscience of right and wrong nor how it drugs them into passivity in the face of practically irresistible power. They do not clearly recognize how it fosters that sense of apathy, meaninglessness ,and despair recently documented as widespread in America.[28] TV and movie showings of this kind often intensify fear in minds already upset and anxious about their urban or suburban environment. They also foster minds whose despair leads them to give up on living genuinely human and humane lives. Little wonder, indeed, that in minds thus saturated few green sprouts of hope and creativity arise. For they have been deluged with a flood of concentrated media violence and sex. How rare is the movie or TV show that presents the real lives of those good ordinary citizens whose routine lives are dedicated to humanizing homes and schools!

Impact on medicine

In medicine the electronic revolution is already achieving wonders. Working from select cells in culture dishes, tissue engineers are already biogenerating skin, bone, and cartilage for clinics. On the draft board are lab-grown organs like heart valves and kidney tissue. Perhaps shortly after 2020 bioengineered eyes and brains will receive AMA approval. As the founder of the bioengineering science, Joseph Vacanti, says, "In the coming century, I think we'll learn to generate living prosthetics for every organ system in the body."[29]

The electron microscope will be greatly improved and thus promote rapid advances in genetic research. Electronically-based body-scanners, with their super-acute eyes will be able to probe into tumors, fetal blockages, and diseased areas of a patient's most intimate organs, even down into the genetic structure. Radiology, already precise in its treatment of certain forms of cancer, will be used in laboratories to create new kinds of treatments. Physicians will need only a computer keyboard to learn about the latest successful advances in Chinese or Russian or Cuban medicine. From central data banks they will receive ultra-precise diagnoses of strange complexes of medical symptoms. They will order prescriptions electronically for patients thousands of miles distant.

Meanwhile, nurses will be freed from the burden of having to write out in longhand a record of treatments given and changes undergone in a patient's condition. Instead, equipped with video recorders and microphones on their lapels, they will create legally acceptable records which physicians can see and hear simply by pressing a button. Already medical secretaries at their computers generate reams of records and bills and process insurance claims. Perhaps they will soon bypass the paper industry and ease mail carriers' burdens. Through the Internet medical secretaries may soon send coded bills and encrypted reports into which patients and insurance firms can gain entry at some publicly available computers while maintaining patients' privacy.

Lastly, the many fruits of these factors of culture lie in the philosophy, morality, religion, and religious groupings which characterize a civilization. How will the media, continually upgraded by such electronic innovations as the foregoing, affect these most precious aspects of national and global cultures?

Impact on philosophy

Something far more than a technological revolution is occurring in people's philosophy. For the media fundamentally reshape the elements through which people both grasp the world about them and verify and express what they grasp.[30] This tremendous power of the media defines not only how people think, but also what they will and will not think about. Moreover, the staccato of mass media's sound-bites and sight-bites impact in a new way on each audience member's consciousness. Impressed this way, these bits tend to work according to psychic blips--that is, neither linearly in step-by-step thinking nor rationally in tracing out consequences. As a result, the media tend to promote spontaneous choices arising from momentary impulse rather than thought-out decisions. Plop choices follow the fast-paced rush-rush consciousness that marches in step with many movie and TV shows and commercials. These choices neither follow upon careful deliberation nor are made when people are in touch with their deeper centers where they can make genuinely discerning decisions.

Furthermore, as studies show, the media control where millions of people will and will not direct their attention. Working like the Coca-Cola logo, a STOP sign, a baseball hat worn backwards, or any other unit of culture--technically called a meme--the media lead millions to value certain things but to regard of little worth other things that the media do not focus on. Since the media rarely attend to the destitute billions of the Third World, its non-attention to these peoples keeps them regarded as of little worth. Meanwhile, the media's continuing focus on the world's relatively few affluent millions tends both to make these latter prefer themselves even more and to prod billions of others to ape them.

Impact on morals

Some try to balance the media's usual lack of concern for the plight of the world's poor by listing the many benefits the media generate. Factually, however, the mass media also often block the enhancement of the human family by their frequent broadcasts of dehumanizing content. This includes their continuous propagation of secularism, consumerism, and materialism.[31] Without overlooking these strong but misleading currents, Americans also need to weigh the different moral impacts produced by the media before and after 1965. The generations of WW I, WW II, and the Korean and Vietnam wars faced common enemies and had cohesive ambitions.[32] Newspapers, radio, and eventually TV revealed tough contacts with harsh realities and helped nurture those generations on great causes. Since 1965, however, Americans have generally lived in prosperous and peaceful times nationally. In these three decades their dreams have come largely via TV and movies. Of course, Americans still tie ribbons round trees and wear lapel buttons, but the epic-size causes are gone and in their place "every news story becomes a crisis, and heroes come cheap," as columnist Jon Meacham writes. He adds, "Multimedia experience is pretty weak compared with what came before us [before 1965]. . . . The cumulative effect of such confected moments [by the media]--and the media's appetite for instant drama never helps--is to trivialize almost everything."[33] This flattening out of ideals, this loss of causes, and this settling down into the yen for usually being comfortable, interspersed with occasionally exciting flings of highly pleasurable escapes, have impacted the soul of Americans.

Media leaders determine policies, strategies, and programs for media communication. Constrained by their ad-driven system, these leaders often know and desire better things to communicate, but find themselves caught in the system. They often seem amoral and quite limited in their mental grasp of the in-depth moral problems just described as well as of those that underlie the promotion of a global culture that genuinely benefits the whole human family. Although media leaders heavily influence the morals of millions, they are not held internationally accountable by the U.N. for either their own character and its development, or their moral/amoral attitudes or their programming policies. Furthermore, for financial reasons, many mass media personnel, pressured at work to the point of burn-out and yet challenged to focus on the positive, fail to be original and creative enough to generate high quality material for dissemination. Thanks to the stimulus provide by the recently founded *American News Service*, some remedies for these ills may soon be generated by Americans practicing democracy locally. This Service, a pioneer in solution-oriented journalism, disseminates its quarterly, *Doing Democracy*, which carries stories of citizens' innovative solutions to locally pressing problems--solutions that can be imitated nationwide.[34]

Finally, both media leaders and their rank and file members often forget the unbridgeable limits of the mass media. For the most part, moral relations are generated directly in and through immediate interpersonal encounters with persons closest to oneself. Towards this goal, the media succeed in stimulating some interpersonal communication by programming dialogue-groups and discussion of films and broadcasts. Nevertheless, since the media are directed to a mass audience, they cannot take the place of those immediate, in-depth, interpersonal contacts and interactions that occur between family members and close friends to give birth and nurture to the moral life.

Impact on religion

How does the media revolution impact on churches, synagogue and mosques? Despite recent scandals, the electronic churches of Christian evangelicals seem not only to have recouped their mass appeal but also to have grown into a multi-billion dollar industry. In the mid-90's, the media-based power of the Christian Coalition put a new kind of fear of God into politicians and their parties. Video and audio tapes now allow people to participate in contemplative exercises according to Zen, healing music, fantasies suggestive of dreams, stories of conversion and papal Masses. Families sitting before a television set can be carried into the personalized intimacy of artists skillful in Gospel story-telling.

If in the future the Vatican Radio were to be elevated into an internationally available TV network, anyone in the world with a TV set could invite the Pope into his or her living room. In religiously affiliated universities, pastoral psychologists will surely design multi-media programs that both respect the individual student's freedom of choice even while also disposing him or her toward an authentic moral and religious conversion. How sharply this would contrast with TV's present tendency to foster in its audience a sense of a rootless superficial functionalism and an ultimate meaninglessness.[35]

Impact on religious groups

Turning to what may be called the media's ecclesial aspect, one can finally inquire how the media will affect religious groupings. Many signals witness to the inroads which the electronic age has already made in organized religious circles. Television evangelists have their "electronic churches." Bibles, hymnals, and church bulletins are produced by computer. Masses and festive celebrations are televised. Bishops and ministers fax messages to each other and engage in satellite-mediated audio or video-conferencing. Young ministers have their preaching or homilizing recorded by TV cameras so that through later critiques they can enhance their ministry of the Word to the faithful.

Moreover, the media will increasingly affect how peoples of the church, synagogue, and mosque are to interact. For the ongoing revolution in communications tends to transform the proposed models for these people's acceptable attitudes and behaviors--for instance, in carrying on respectful dialogue, maintaining polite behavior when hearing persons of different opinions, being open-minded and open-hearted to peoples of vastly different cultures. Unfortunately, injustices towards women, blacks, and other minorities, even though somewhat reduced, perdure into the present as proofs of the sinfulness of these groups' members and usually in spite of their official policies. Yet, if most members come to acknowledge these past and present injustices, it will inject a new sensitivity and spirit into meetings where former power-holders are now called to interact with previously powerless members.

This chapter, then, has attempted the impossible task of predicting the media scene in the year 2020. Some of the future, seen through our gazing into the crystal ball of cyberspace, seems highly probable, but other portions of the vision may prove to be simply unfulfilled prophecy.

In conclusion, then, is American culture stronger because of the overall impact of the media? When members of the media's PR department reactively reply with a resounding "yes," their saying so does not make it so. Rather, the pressing need is to examine carefully what new attitudes and behaviors can more appropriately help us meet this in-rushing cyber-age. For this, each person needs to develop his or her own response. Such a response needs to embrace the best values of our evolving past--both our American and our global past. It must also embrace humanity's highest hopes for the future with a new, more effective and communally responsible attitude enhancing the future. Such is the task of the next and final chapter.

QUESTIONS FOR DISCUSSION

1. As you consider the trend of changes in the economy and in finance, what signs of hope for the future do you see?

2. What feelings does the prospect of cyber-war arouse in you? Do you see this possibility as motivation to work for peace?

3. Should the U.N. try to exert some real but limited control over the electronic media, including Internet and World Wide Web? Why or why not?

4. Do you believe there is a movement toward a common worldwide language? What would be some of the positive results? Would there be any negative effects?

5. How will the new, more powerful electronic technologies benefit science and medicine?

6. Are you concerned about any of the long term psychological effects of the mass media? What specific effects cause you concern?

7. Do you believe that Americans have developed religiously and morally through use of the mass media? Why or why not?

Chapter 11

MANAGING THE MEDIA MAZE

In response to the preceding chapters, the reader may feel challenged to take some steps to manage the media maze. What, then, can people do to gain greater control of their lives as they use the various media? This final chapter presents some suggestions for certain positive attitudes and specific actions which individuals and groups can adopt.

The reader will recall from Chapter 2 the media's many good aspects which grounded the adoption of a primarily positive attitude toward the media. Moreover, the reader has a right to expect reasonable help from the media to become a better person, a more enlightened citizen, an individual more relaxed and refreshed at the end of a day's work. Consequently, the person who uses the media has a right to a choice of the kind of programs that meet these expectations.

Adopting a New Attitude toward Communication

Communication can no longer be considered primarily as a transmission of messages but rather as a ritual participation. A communications expert, James Carey, explains it this way, "A ritual view of communication is directed not toward the extension of messages in space but toward the maintenance of society in time; not the act of imparting information but the representation of shared beliefs."[1] Shared beliefs form the roots of the democratic social order. The media, however, play a leading role in the change of this social order which is underway in the United States today.

The outdated, but still influential, transmission view of communications involves its holder in a model that requires power in the sender. It also evokes anxiety in the receivers who sense the sender's aim to bring their minds into conformity. By contrast, the ritual view requires participation together, sharing in the co-creation of feeling and life, of ideals and meanings. Moreover, it fosters mutual enrichment out of respect for, and understanding of, diverse positions. A primarily ritual view of TV and other media, implies less attention to the information being sent and more attention to which particular view of the world is portrayed and confirmed.[2] A ritualist viewer refuses to go along with the media's usual view that language is "an instrument to manipulate objects. . . [and] to get others to believe what we want them to believe." Rather he or she regards the media as "a device to establish the truth" which strengthens and increases genuine community.[3] In brief, as Professor Carey comments:

> We must first discard the view of language as reference, correspondence, and representation and the parallel view that the function of language is primarily to express assertions about the world. Then we must substitute the view that language--communication--is a form of action--or, better, interaction--that not merely represents or describes but actually molds or constitutes the world.[4]

To bring about this change of view, however, requires that the person be on guard against that "nostalgia of the future." If dreams of future possibilities distract people, they are kept from contributing to building the present moment through socially sensitive and creative actions. The glow of the bright future prevents the investment of energies in the present.

Media analyst, Hamid Mowlana, and other communications researchers adopt views that confirm Carey's recommendation.[5] They see a growing desire to define communication primarily in terms neither of the technology that transmits bits of information, nor of the data and knowledge transmitted. Instead, they believe that individuals and some nations are beginning to look at communication as human interaction. Seen in this way, communication can open the way to trust and understanding.

The problem still remains: How stress the primacy of human interaction in communication and thus counter the opposite trend to emphasize its technology? Mowlana suggests two steps:

> 1) We must shift our attention and our emphasis from co.nunications (as means) to communication (as sharing and trust). Intellectually, this will require a reorientation in communication studies: from sole concern with roles, effects, and impacts of communication media to the study and discovery of a communication theory of society.
>
> 2) We must create an environment in the form of a restraining influence that can stop this deterioration in relations, can protect humanity from self-destruction, and can eventually direct the machinery of media- communication to explore human growth and potential.[6]

A condition for escaping the anxiety created by a transmission view of communication and for adopting the recommended new view of communication lies, according to Hamid Mowlana, in a rethinking of education. Without this, few will actually view communication as participation. In a statement deserving protracted quotation Mowlana states:

> The general tendency now is to emphasize the vocational aspect of education, using a hard-headed, businesslike tone. Higher education is represented as an industry engaged in manufacturing socially needed commodities such as secretaries, engineers, and even communicators. As usual, the central claim is efficiency, and the logic is that mass education requires mass-production methods. To some critics, the result is already clear: "technication" meaning standardization. In the age of cybernetics and our highly technological society, the process of personal orientation and self-discovery is becoming not a luxury but a necessity. On most of our college campuses few students, apart from those who have a vocation and know exactly where they are going, know enough either of themselves or of the world to make effective choices. . . . Education is the mainstay of culture, for how one learns is culturally determined; but flexibility and creativity are the keys to positive change and growth.[7]

More specifically, our schools can help develop in students a critical distance towards TV and other media. In school students should be taught TV/media literacy. They should be taught how to distinguish true and misleading statements and how to recognize cheap manipulation.[8] Already educators, confronted with the computer, have become somewhat conscious of the effects of media on consciousness and on the educational process. The real question, which educators have not yet faced squarely is, "How can we use education to control TV (or the computer or word processor)?"[9] Schools should teach the young how to interpret the symbolic forms of their culture, and surely in our culture TV is a leading symbolic form. Hence, as Neil Postman concludes, schools should "require that they [the students] learn how to distance themselves from their forms of information. . . . [This] is not so bizarre an enterprise that we cannot hope for its inclusion in the curriculum; even hope that it will be placed at the center of education."[10]

Specific Individual Attitudes and Actions

The above mentioned foundational attitude toward communication is needed to free receivers from being dominated by the media. Other more specific attitudes and actions will help towards the same end and will be examined individually and then socially.

A first specific attitude is "Be Selective!" Media users need to sharpen their attention and to select media and products more discriminately. In his book, *Advertising*, Lester G. Telser wrote about selectivity, "In modern industrial society consumers can choose from an ever-increasing and ever-changing array of goods. Wise consumption is difficult under these conditions and demands a degree of skill not required of shoppers in a simpler economy."[11] To be attentive and selective when using TV, movies, videos, popular books, and magazines, it is necessary to:

1. Choose only the best in each of these media fields. Do not waste time on cheap and mediocre goods. For example, watch for quality programs on TV that are stimulating and educational as well as entertaining--shows like the Civil War series, Masterpiece Theater, some documentaries on PBS or offerings by *National Geographic.*

2 Listen to the advice of one shrewd appraiser of TV: "If you want relief from long hours of work without becoming hooked into passive viewing, a very brief bit of TV can refresh your mind. If you watch TV too long, you will become a passive spectator and neither refresh your mind nor invest your time in a worthwhile way." Remember when a tired person sits before the TV set, the body seems to relax while the mind goes into neutral--into a semi-dazed state where little is absorbed consciously. Under such a

condition, it is difficult to stop watching TV.

3. "If you doubt, don't watch." This rule of thumb, by Dr. Eric Anderson, media specialist at Cincinnati's Xavier University, directs a person how to act when in doubt about the real value of a program.

4. Avoid plopping down a restless, crying child before the TV set so that you can do some work! As a parent or one responsible for young children, you are familiar with this challenging temptation. Yet you are aware of the effect of most TV programs on a young child. Pause to find other options to distract the restless child. For example, give the child a picture book from the library. The more children read, the less they watch TV. Parents must take control of TV and let children watch specific <u>programs</u>, not <u>TV</u>. The V-chip and program ratings can help towards controlling what children watch.[12]

5. Use differently the two hours you spend weekly turning over pages of ads in the magazines or newspapers you ordinarily read. Instead, use the Internet or the library to contact higher quality newspapers like the *New York Times*, the London *Economist*, or some good Indian newspaper written in English. If you read a foreign language, reach out for the best foreign newspaper or magazine and discover its kind of reporting and editorial approach. It will likely be quite different from the tone and slant of our American press. Your stretching to other sources will give you more breath for evaluating the perspectives commonly assumed by American media.

6. Be selective in choosing from the books flooding the market each year. Tell yourself, "I have time only to read the very best books!" Be selective, too, about movies and videos. Just because the latter are economical does not mean that most of them are worth watching. Yet the availability of video classics makes it possible for parents to watch and discuss them with their children.

7. Be hawk-eyed when reading newspapers. Watch for phrases like "it's alleged", "reportedly," "reputed," and "according to informed sources [not identified]." This attitude can keep one from uncritically swallowing statements that are not based on fact.

Specific Social and Political Attitudes and Actions

The individual attitudes and actions just mentioned need to be supported by social and political ones. For example:

1. Organize a TV quality-control group in your neighborhood. The members of such a group commit themselves to watching only the highest quality programs and to informing others of their choices. As helps towards this, subscribe to *Connect*, the journal of the Center for Media and Values, and consider the solutions on managing made available though the *American News Service*.

2. Interact with people who rarely view TV and contrast their ways of dealing with human relations and the world over against the media-filtered ways of most Americans.

3. Insist that media directors present an accurate image of other cultures with their unique values and ways of life rather than imposing American values on Latin Americans, Africans, Chinese, etc. A gathering of letters or e-mail messages is effective here.

4. Call for accountability from the corporate giants. The mass media are too important to the democratic process to allow them any complete one-sided control over the information that reaches every segment of our society.

5. Encourage the media executives to include in their broadcasts a fairer representation of the common religious values of the world's religious peoples. No need to make secular humanism the politically correct language in broadcasting.

6. Initiate the creation of a consumer group of concerned neighbors who can conjointly boycott inadequate programs, especially those showing violence simply to grab attention. A simultaneously timed switch-off of as many TV sets as possible in a particular city will alert media directors to the public's dissatisfaction. Advertisers will pressure the media to change the programs or lose advertising revenue.

7. Use interactive media when available to register your choices with producers. Choices of the audience send messages to producers who wish to stay economically viable.

8. Write editors demanding a shift in their assignment of investigative reporters. The current ratio of sending only 6 out of 1,110 investigative reporters to uncover white-collar crime in corporations is scandalous.[13] When writing editors, press them to resist the itch toward sensationalism and to assign more investigators to corporate crime.

9. Seize every opportunity to learn more about Third World cultures. A visit to a Third World country--not as a tourist but as one helping those people develop themselves—will reward you with a different perspective for evaluating the ideology spread by our American media. On your return, you will be struck by the contrast between our individualism, consumerism, and secularism and the attitudes and behavior of Third World peoples.

In his book, *The Geopolitics of Information*, Anthony Smith sketches the background for a sound appraisal of our system of international communications. In connection with much current discussion of a "new industrial revolution" made possible by electronics and satellites, Smith writes:

> Preparations for a new industrial revolution on such a scale as that made necessary by the advent of the new micro-electronics and the new telecommunications systems should be conducted more with a view to its global implications than to the short-term interests of the transnational companies and quasi-monopolies which dominate these fields. . . .The new media have the power to penetrate more deeply into a "receiving" culture than any previous manifestation of Western technology. The results could be immense havoc, an intensification of the social contradictions within developing societies today. In the West we have come to think of the 2,500 communication satellites which presently circle the earth as distributors of information. For many societies they may become pipettes through which the data which confers sovereignty upon a society is extracted for processing in some remote place.[14]

Accordingly, we need to adopt a global view. In the media field, it is becoming increasingly short-sighted to take a restrictive national view. Only a global viewpoint is appropriate for a technology with a worldwide influence.

10. Employ ethical principles that promote the dignity of every human person. This is the foundation for the shift from a manipulative style of communication to one that puts greater stress on the interaction of freely responding persons, who also reach out socially and politically.

Toward embodying these ethical principles, Mowlana proposes four basic norms and some suggested ways of implementing them:

I) Prevention of war and promotion of peace:
(a) by increasing information on peaceful solutions to conflict,
(b) by breaking down stereotypes that dehumanize populations,
(c) by conferring prestige on the peacemaker,
(d) by helping to create a spirit of reconciliation, and
(e) by putting peacemakers on opposite sides in touch with one another.

II) Respect for culture, tradition and values:
(a) by promoting respect and tolerance for the world's manifold cultures,
(b) by facilitating the communication between cultures,
(c) by helping diverse value systems arrive at common definitions for universal goals as peace, integrity, and national sovereignty,
(d) by strengthening and preserving cultural identities.

III) Promotion of human rights and dignity:
(a) by publicizing violations of human rights,
(b) by promoting the democratization of communication.

IV) Preservation of the home, human association, family and community:
(a) by promoting interpersonal communication by facilitating more interaction among people rather than narcotizing them through mass-distributed programming, and
(b) by facilitating self-reliance and interdependence by publicizing local decentralized solutions to common problems.[15]

Conclusion

Throughout this book we have seen that today, in this exciting age, we can expect marvelous results from new and powerful technology. In a democratic society, the wise use of such power must not be left to government, big business, or a few media directors. Every American shares the responsibility for the choices that will affect our nation and the globe.

This last chapter suggests some ways of carrying out that responsibility. The suggested steps do not constitute an exhaustive list of possibilities. You may think of other ways that are more practical for you. You are invited to join with other concerned Americans in action that can lead us to greater freedom in our personal lives, strengthen the image of the United States as a leader in world peace, and bring us into the shared freedom of participatory communication.

QUESTIONS FOR DISCUSSION

1. In all you have read here and elsewhere, what do you find most motivating to move you to action in regard to the media, especially TV?

2. As an individual, what can you do to improve the quality of TV programs?

3. What action can you take with others to influence the media?

ENDNOTES

Introduction

1. Kenneth Gergen, *The Saturated Self* (NewYork: Basic Books, 1991).

Chapter 1

OVERVIEW OF THE MEDIA TODAY

1. *The New York Public Library Desk Reference*, 2nd edit. (New York: Prentice-Hall, 1989), 97-101.

2. W. Russell Neuman, *The Future of the Mass Audience* (New York: Cambridge University Press, 1991), 116.

3. Ibid. 44.

4. Alfred N. Whitehead, *Science in the Modern World* (New York: Free Press, 1967), 96.

5. *The New York Public Library Desk Reference*, 101-102.

6. Neuman, 44.

7. James Reston, *Deadline* (New York: Random House, 1991), 468.

8. Doug Underwood, "Identity Crisis of Newspapers" *Impact of Mass Media: Current Issues*, ed. Ray Eldon Hiebert, (White Plains, New York: Longman Publishers USA, 1995), 36.

9. Editorial, "The End of Mass Media?" *Futurist,* January, 1994.

10. Ken Auletta, *Three Blind Mice: How the TV Networks Lost Their Way* (New York: Random House, 1991), 570.

11. Ibid.

12. Gregor T. Goethals, "TV's Iconic Imagery in A Secular Society," *New Theology Digest* 6 (February 1993): 40.

13. Neuman, 91.

14. Neuman quotes studies by Naples, (1979), Lodish, (1986), and McGuire, (1986).

15. Neuman, 94-5.

16. Ibid., 95.

17. Ibid., 104-5.

18. Mario A. Orlandi, Cynthia E. Dozier, and Marlene A. Marta, "Computer-Assister Strategies for Substance Abuse Prevention: Opportunities and Barriers," *Journal of Consulting and Clinical Psychology* 58 (1990): 425-31.

19. Neuman, 106-7.

20. Ibid., 165.

21. Ibid.,166.

22. Andrew Lippmann as quoted by George Gilder, *Life After Television* (New York: W. W. Norton, 1992), 123.

23. Alvin Toffler, *Powershift* (New York: Bantam Books, 1990), 360-4.

24. *Statesman Year, 1996-1999*, 133rd ed., (New York: St Martin's, 1996)

25. Toffler, 363.

26. Ibid., 364.

27. Ibid., 470.

Chapter 2

CHOICE OF A POSITIVE APPROACH TOWARD TELEVISION

1. Jesus in the Sermon on the Mount, Matthew 7: 3.

2. For example: James W. Carey, *Communication as Culture* (Boston: Unwin Hyman, 1989).
John A. Coleman and Midlos Tomka, edd., "*Mass Media*" in *Concilium 1993/6* (Maryknoll, NY Orbis, 1993).
Wilson P. Dizard, Jr., *The Coming Information Age*, 3rd edit., (New York: Longman, 1989).
Joseph R. Dominick, *The Dynamics of Mass Communication*, 5th edit., (New York: McGraw-Hill, 1996).
G. F. Jankowski and D. C. Fuch, *Television Today and Tomorrow: It Won't Be What You Think* (New York: Oxford University Press, 1995).
Joli Jensen, *Redeeming Modernity:Contradictions in Media Criticism* (London: Sage, 1990).
Walter Ong, S.J., *Orality and Literacy* (London: Methuen, 1982).
David L. Altheide, *Media Power*, vol. 158 of the Sage Library of Social Research, (Beverly Hills: Sage, 1985).

3. The American Psychological Association Task Force, *Big World, Small Screen: The Role of Television in American Society,* edd., Aletha Huston et al., (Lincoln, NE: University of Nebraska Press, 1992); [hereafter *BWSS*].

4. See n. 2 above.

5. Ong, *op. cit., passim*

6. *Time,* 146 (July 12, 1995).

7. These groups were the targets selected by the American Psychological Association Task Force for its 1992 research, as published in *BWSS.*

8. Neil Postman, *Conscientious Objections* (New York: Knopf, 1988), 66.

9. Ibid., 18-9.

10. *BWSS,* 83.

11. Ibid., 63-5.

12. P. Kelley and B. Bunter, "Helping Viewers Learn from Television," *Journal of Educational Media* 22 (March 1996): 23-40.

13. S. Marriott, "Time and Time Again: 'Live' Television Commentary and the Construction of replay Talk," *Media, Culture & Society,* 18 (January 1996: 69-86.

14. Charles Van Doren, *A History of Knowledge Past, Present and Future* (New York: Carol Publishing Co., 1991: 373.

Chapter 3

TRUTH IN ADVERTISING

1. Mark Crispin, *"Free the Media,"* Nation 262 (June 3, 1996): 12.

2. Courtland L. Bovee et. al. *Advertising Excellence,* (New York: McGraw Hill, 1995) 66.

3. "FTC to Media: Screen Ads Better," *Advertising Age* 67 (January 15,,1996).

4. *Consumers' Research Magazine,* 79 (June 1, 1996).

5. Harry James Cargar, "America is Saturated by Violence," *Impact of Mass Media: Current Issues,* ed. Ray Eldon Hiebert, (White Pains, NY: Longman Publishers USA, 1995), 205.

6. William Meyers, *The Image-Maker* (New York: The New York Times Book Co., 1984), 15 ff.

7. Ibid,. 56.

8. Vance Packard, *the Hidden Persuaders* (New York: David McKay, 1957), 37.

9. "The Rebate that Isn't," *Consumer Reports* 51 (January 1986): 67.

10. June Fletcher, "The 7 Deadly Sins," *Builder* 13 (September, 1990): 64.

11. "Informercials under Fire," *Consumer Research* 73 (September, 1990): 2.

12. Bovee et al., 396.

13. Jean Kilbourne, “Deadly Persuasion,” in Hiebert, 211-3.

14. Philip Elmer-Dewitt, Hooked Up to the Max,” *Time,* 244 (September 26,1994): 58-60.

15. Mallory Woeber and Barrie Gunter, *Television and Social Control,* (New York: St Martin’s, 1988), 205.

16. *Consumer's Report* 51 (June 1986): 243.

17. Terrance H. Witkowski, “Self-Regulation will Suppress direct Marketing’s Dark Side,” *Marketing News* 23 (April 24, 1989): 4.

18. *Consumer's Report* 51 (June, 1986): 423.

19. “Mail-Ordering Shopping,” *Consumer's Report* 59 (October 1994).

20. Benjamin H. Bagdikian, *The Media Monopoly*, 4th edit., (Boston: Beacon, 1992), 249.

21. Rachael M. Lauer, "Evaluating Sex in Advertising," *Humanist* 48 (May/June 1988): 25.

22. Cyndee Miller, "We've been Cosbyized," (Cosby Show) *Marketing News* 24 (April 16, 1990): 1-2.

23. Andrew Sullivan, "Flogging Underwear," *New Republic* 198 (January 18, 1988): 20-4.

24. Alice Z. Cuneo, "Ads Put Idle Computers to Work," *Advertising Age*, 65 (December 8, 1994): 23.

25. David Pescovitz, "The Future of Advertising," *Wired,* 4 (April 1996): 68.

26. David Kalish, "Creative Concepts--Now You Hear It," *Marketing and Media Decisions* 23 (May, 1988): 32.

27. Sidney Hecker & David W. Stewart, edd., *Nonverbal Communication in Advertising* (Lexington, MA: D.C. Heath and Co., 1988), 248; (emphasis added).

28. Ibid.

29. Joe Saltzman, "Commercials Defy Logic," *USA Today* 16 (January 1988): 93.

30. Rhoda H. Karpatkin and Anita Holmes, "Making Schools Ad-Free Zones," *Educational Leadership,* 52 (September 1995): 72-6.

31. William J. Donnelly, *The Confetti Generation* (New York: Henry Holt, 1986) 201.

Chapter 4

SELECTION AND BALANCE OF THE NEWS

1. Goethals, 47.

2. Bagdikian, 13-26.

3. Goethals, 48.

4. J. McManus, "A Market-based Model of News Production," *Communication Theory* 5 (November 1995): 301-38.

5. Bagdikian, 27-45.

6. Anthony Smith, *The Geopolitics of Information: How Western Culture Dominates the World* (New York: Oxford, 1980), 232-3.

7. Andrew Kopking, "The Unwritten Watergate Story," *More* 4 (November 1974).

8. Landrum R. Bolling, ed., *Reporters Under Fire: U.S. Media Coverage of Conflicts in Lebanon and Central America* (Boulder, CO: Westview Press,. 1985), 13.

9. Altheide, 45.

10. Ibid., 128-9.

11. David Harvey, *The Condition of Postmodernity* (London:Blackwell, 1989), 159.

12. Summary of F. Fico and S. Soffin, "Fairness and Balance of Selected Newspaper Coverage of Controversial National, State, and Local Issues," *Journalism and Mass Communication Quarterly* 72 (Autumn 1995): 621-33 as reported in # 302 of *Communication Abstracts* 19 (1996): 172.

13. L Steiner, "Judging the Judges: Journalists and Standpoints, " *New Jersey Journal of Communication* 3 (Fall 1995): 133-50.

14. For the differences worked through television when news is framed episodically, rather than thematically, see Shanto Iyengar, *Is Anyone Responsible? How Telvision Frames Political Issues* (Chicago: University of Chicago Press, 1991), 2-3, 143, 163, n.19. Iyengar documents how framing news items episodically, rather than thematically, keeps the audience from criticizing those responsible for injurious events and thus favors the *status quo.*

15. Ibid., 5, 140-43.

16, James Martin, "You shall Not Oppress an Alien," *America* 175 (December 14,1996): 25.

17. Bolling, 142. Dr. Gomez focused only on European sources: Reuters in England, AFP (Agence France Presse) in France, EFE (European Free Trade Bulletin) the European Union, and ANSA (Agenzia Nationale Stampa Associata) in Italy. In addition, the news wire services--IPS (International Publishing Services) of non-aligned nations and SalPress of Latin America--are employed only minimally by most U.S. newspapers and TV networks.

18. Jack Nelson-Pallmeyer, *War Against the Poor* (Maryknoll, NY: Orbis, 1990), 62.

19. William Preston, Jr., et al., *Hope and Folly: The United States and UNESCO, 1945-1985*(Minneapolis: University of Minnesota Press, 1987), 214-5.

20. Bolling, 144.

21. Shanto Iyengar, Mark D. Peters, and Donald R. Kinder, in *Media Power in Politics*, ed. Doris A Braber of the University of Illinois at Chicago, 2nd edit., (Washington, D.C.: Congressional Quarterly Press, 1990), 55-8.

22. Charles W. Ostrom, Jr. and Dennis M. Simon, "The Man in the Teflon Suit?" Public Opinion *Quarterly* 53 (1989): 353-87.

23. On June 7, 1984, in its program, "All Things Considered," National Public Radio aired a report on the kinds of music used as background for news stories.

24. *National Jesuit News* 26 (December 1996-January 1997): 2.

25. K. Ross, "Gender and Party Politics: How the Press Reported the Labour Leadership Campaign 1994," *Media, Culture & Society* 17 (July 1995): 499-509.

26. *BWSS*, 28.

27. Ibid., 133.

28. J . D. Robinson and T. Skill, "The Invisible Generation: Portrayals of the Elderly on Prime Time ," *Communication Reports* 8 (Summer 1995): 111-9.

Chapter 5

ACCURACY AND CLARITY OF THE NEWS

1. George Juergens, *News from the White House* (Chicago: The University of Chicago Press, 981), 6-8.

2. Ibid., 40.

3. Ibid., 43.

4. James Deakin, *Straight Stuff* (New York: William Morrow, 1984), 82.

5. David Chagall, *The New Kingmakers* (New York: Harcourt Brace Jovanovich, 1981), 301.

6. Deakin, 83.

7. Joyce Nelson, *The Perfect Machine* (Philadelphia, PA: New Society Publishers, 1992), 110.

8. Deakin, 85.

9. Goethals, 44-5.

10. Jim Willis, *The Age of Multimedia and Turbonews*, (Westport, CT: Praeger, 1994), 180-1.

11. Deakin, 95.

12. Ibid., 98-9.

13. J. D. Robinson and T. Skill, 111-9.

14. The Committee of Santa Fe, "A New Inter-American Policy for the Eighties," (Washington, D.C.: Council for Inter-American Security, 1980), 1.

15. Nelson-Pallmeyer, 47.

16. *Newsletter of Middle East Action Committee of Quixote,* (Hyattsville, MD: 1980), 1.

17. William J. Taylor, Jr., and James Blackwell, "The Ground War in the Gulf, " *Survival* 23 (May/June, 1996): 234.

18. Douglas Kellner, *the Persian Gulf TV War* (Boulder: Westview Press, 1992), 67-8.

19. Jean Baudrillard, The Gulf War did not take place, (Bloomington, IN: Indiana University Press, 1995), 13-4.

20. *Newsweek* 117 (March 18, 1991): 38.

21. Goethals, 49.

22. Ibid. 50.

23. Debra Hernandez, "Eyes in the Sky," *Editor & Publisher,* 129 (April 27, 1996):52-3.

24. Willis, 164.

25. Margaret Carlson, "New Lows for News," *Time* 148 (July 15, 1996): 12.

26. Andrew O. Shapiro, *Media Access* (Boston: Little, Brown and Co., 1976), 15-6.

27. Ibid., 18.

28. Shanto Iyengar, *Is Anyone Responsible?* 2-3, 143, 163n19. See above 201n14.

29. Ibid.

30. Doris A. Graber, *Mass Media and American Politics* ,3rd edit., (Washington, D.C.: CQ Press, 1989).

31. Richard Schickel, "How TV Failed to Get the Real Picture," *Time* 139 (May 11, 1992): 29.

32. Ibid.

33. Goethals, 47.

34. Shapiro, 20-4.

35. Ibid., 18.

6. *Many Voices, One World: Communication and Society, Today and Tomorrow, The McBride Report* (Paris: UNESCO, 1984), 36.

37. Garth S. Jowett and Victoria O'Donnell, *Propaganda and Persuasion* (Beverly Hills: Sage, 1986), 92-3.

38. Smith, 66-7.

Chapter 6

AVAILABILITY OF INFORMATION

1. Neil Postman, *Amusing Ourselves to Death* (New York: Penguin Books, 1985), 68.

2. Ibid., 65.

3. Ibid., 67.

4. Bianca P. Floyd, "Program in Afro-American Studies Explores the Racial Gap in Access to Technology," *Chronicle of Higher Education* 43 (December 20, 1996): A 19-20.

5. Juergen Arthur Heise, *Minimum Disclosure--How the Pentagon Manipulates the News* (New York: W. W. Norton, 1974)

6. Joyce Nelson, 134 ff.

7. *U.S. Public Law* 89-554.

8. Ibid., (emphasis added).

9. "Credit for Reno," *Editor & Publisher* 128 (September. 30, 1995): 6.

10. Debra Gersh Hernandez, 'Dealing With The FoI Crunch," *Editor & Publisher* 129 (August 3, 1996): 14.

11. *American Libraries,* 7 (September 1979): 223-5.

12. *American Libraries*, 18 (March 1987): 217.

13. "Less Access to Less Information by and about the U.S. Government,

14. Ibid.

15. Elie Abel, "Looking Ahead from the Twentieth Century," in *Issues in New Information Technology*, ed., Benjamin M. Compaine, (Norwood, NJ: Ablex Publishing Corporation, 1988), 180.

16. *ALA Washington Office Chronology*, 16: 1.

17. Michael Fox, "Siege of the 1st Amendment," in *Impact of the Media*, ed. Hiebert, 94.

18. Ralph Nader, "Don't Let the Sunshine In," The Nation 233, (November 1981): 470-3.

19. Hernandez, 15.

20. Ibid.

21. James Ridgeway, *Village Voice*, cited in *ALA Washington Office Chronology*, 16: 11.

22. *Morrow's International Dictionary of Contemporary Quotations* (New York: William Morrow, 1982), 319.

23. Anthony Marro, "When the Government Tells Lies," *Columbia Journalism Review* 23 (March/April, 1985): 30.

24. Ibid., 32.

25. Ibid.

26. Bagdikian, 3rd edit. (1990), 44-5.

27. Kurt Rudolf Mirow and Harry Maurer, *Webs of Power* (Boston: Houghton Mifflin, 1982), 231-2.

28. Ibid., 232.

29. Ralph Nader, 470-3.

30. Mirow and Maurer, 233-5.

31. Rosalie Bertell, *No Immediate Danger*? (Toronto: Women's Educational Press, 1985), 138.

32. Dr. Helen Caldicott, *Missile Envy,* revised edit., (New York: Bantam Books, 1986), 284.

33. Ibid., 290-1.

34. Ibid., 292.

35. Ibid.

36. Elizabeth Neus, "UC and P&G could equal R&D heaven," *Cincinnati Enquirer* (June 20, 1991): B1.

37. Ibid.

38. Ibid.

39. Herbert T. Schindler, "Information--a Shrinking Resource," *The Nation*, 241 (December 28, 1985), 708.

40. Ibid.

41. Ibid.

42. Meredith W. Mendes, "Privacy and Computer-based Information Systems," in *Issues in New Information Technology*, ed. Benjamin M. Compaine,, 210.

43. Jeffrey Rothfeder, "Is Nothing Private?" *Business Week*, # 3,122 (September 4, 1989): 74-82.

44. "Cell-Phones Fraud Bust, " *Consumer's Research Magazine* 8 (August 1, 1996).: 38.

45. News Service America Online, November 15, 1996.

46. Rothfeder, 74-82.

47. Ibid.

48. Anne Wilette, "Local man says service helps clients compete" *Cincinnati Enquirer* (June 17, 1991): A 4.

49. Dennis Camire and Anne Wilette, "Data brokers flirting with privacy issue," *Cincinnati Enquirer* (June 17, 1991): A1- A4.

50. Meredith W. Mendes, "An Update," *Computers and People*, 28 (August 1979): 31.

51. Mendes, 207-208.

52. Wilson P. Dizard, Jr., *The Coming Information Age,* 3rd edit., (New York: Longman, 1989), 204-5.

53. Ibid.

Chapter 7

ROLE OF GOVERNMENT, BUSINESS AND THE MEDIA HIERARCHY

1. John Dewey, *The Public and Its Problems*, (New York: Henry Holt., 1927) where Dewey wrote:

> The smoothest road to control of political conduct is by control of opinion. As long as the interests of pecuniary profit are powerful, and a public has not located and identified itself, those who have this interest will have an unresisted motive for tampering with the springs of political action in all that affects them. . . . The gathering and sale of subject-matter having a public import is part of the existing pecuniary system. . . .[and] the assembling and reporting of news would be a very different thing if the genuine interests of reporters were permitted to work freely.

--quoted through *The Philosophy of John Dewey*, ed. John J. McDermott, (Chicago: University of Chicago Press, 1981), 642.

2. Carl J. Friedrich, *The Pathology of Politics: Violence, Betrayal, Corruption, Secrecy, and Propaganda* (New York: Harper & Row, 1972).

3. Altheide adduces three veirfiable theses that undermine the conspiracy interpretation; see his Media *Power,* 68-9.

4. The centerfold of *The Nation* 262 (June 3, 1996) provides a graphic chart of the extensive tentacles of the "four octopuses" of The National Entertainment State"--General Electric, Time Warner, Disney/Cap cities, and Westinghouse--with the caution, on p. 10, that "this chart leaves out other giants: Rupert Murdoch's News Corporation, John Malone's Tele-Communications Inc., and Sumner Redstone's Viacom," along with "the octopus that is S .I. Newhouse" and foreign players, like Sony (Columbia, Tri-Star)."

5. Here and in the next three pages most of the data is derived from *Broadcast & Cable Yearbook, 1995* (New Providence, N.J.: R. R. Bowker, 1995), vol. 1.

6. John Greenwald, "Rupert's World," *Time* 142 (September 12, 1993), 64-5.

7. *Nation* 262 (June 3, 1996): 15.

8. "A Free and Responsible Press: A General Report on Mass Communication," Commission on Freedom of the Press, ed. Robert D. Leigh, (Chicago: University of Chicago Press, 1947), 61-2.

9. Michael Parenti, *Inventing Reality: The Politics of the Mass Media,* (New York: St. Martin's Press, 1986), 31. For Dewey's warning, see above, 208 n1.

10. Mark Crispin Miller, "Free the Media" *Nation* 262 (June 3, 1996), 14.

11. *TV Quarterly* 28 (1996): 37.

12. Miller, 10.

13. Ibid.,12.

14. Sylvia M. Chan-Olmsted, "From Sesame Street to Wall Street: An Analysis of Market Competition in Commercial Children's Televisionn," *Journal of Broadcasting & Electronic Media* 40 (Winter 1996): 30-40.

15. J. McManus, "A Market-based Model of News Production," *Communication Theory* 5 (November 1995): 301-338.

16. Miller, 12.

17. Schindler, 708..

18. Miller, 10.

19. Debra Gersh, "Governments worldwide continue to interfere with press freedom," Editor & Publisher 122 (February 18, 1989): 16-7, 35-40.

20. Miller, 10.

21. Nelson-Pallmeyer, 62.

22 . Ibid.

23. *Extra*, I, (October-November, 1987): 1.

24. Parenti, 232-3.

25. Nelson-Pallmeyer, 66.

26. Ibid.

27. Miami Herald (July 28, 1988).

28. Nelson-Pallmeyer, 66.

29. James S. Torrens, "U.C.A.--the Witnesses Talk," *America* 163 (November 24, 1990): 394-400.

30. Once the story broke, the *Boston Globe* responsibly kept this item before the public, almost single-handledly among major newspapers.

31. Herbert Mitgang, "Annals of Government," *The New Yorker*, October 5, 1987, 47 ff.

32. Nelson Blackstock, *Cointelpro* (New York: Vintage, 1976).

33. Bernadine A. Hoduski, "The `Lacy Report' and libraries," *Library Journal* 112 (January 1987): 60, and Tess Chichioco, "Making it hard to get records," *Editor & Publisher* 123 (March 31, 1990): 16.

34. George Garneau, "In pursuit of access," Editor & Publisher, (December 24, 1988) 8.

35. Ray Bonner, *Weakness and Deceit: U.S. Policy and El Salvador* (New York Times Books, 1984); also Nelson-Pallmeyer, 64.

36. Noam Chomsky, *Necessary Illusions: Thought Control in Democratic Societies* (Boston: South End Press, 1989), chap. 2.

37. David Harvey, *The Condition of Postmodernity* (London: Blackwell, 1989), 160.

Chapter 8

IMPACT OF THE MEDIA ON ITS AUDIENCE

1. See, for instance, D. Gauntlett, *Moving Experiences: Understanding Television's Influences and Effects*, Acamedia Research Monographs, # 13 (London: John Libby, 1995).

2. William James, "What Pragmatism Means," *Pragmatism* (Cambridge, Mass.: Harvard University Press, 1975), 30.

3. Audiences, however, are far less blameworthy than media directors for most media's tilting more toward profit-seeking than toward the other essential goals of broadcasting--promoting the public good through enlightening their audiences and enhancing character development.

4. Walter Cronkite, *A Reporter's Life* (New York: Random House, 1996)

5. Neil Postman, 105-7, 118, 126, 137, 146.

6. Goethals, 44.

7. L. B. Snyder and R. A. Rouse, "The Media Can Have More than an Impersonal Impact: the Case of AIDS Risk Perceptions and Behavior," *Health Communication* 7 (1995): 125-45.

8. R. A. Weaver-Larisey and S. F. Tinkham, "Advertising Message Strategies in U.S. Congressional Campaigns: 1982, 1990," *Journal of Current Issues and Research in Advertising* 18 (Spring 1996): 53-66.

9. J. J. Cronin, "In-home Observations of Commercial Zaping Behavior," *Journal of Current Issues and Research in Advertising* 17 (Fall 1995): 69-75.

10. F. P. Kasoma, "The Role of the Independent Media in Africa's Change

to Democracy," *Media Culture & Society* 17 (October 1995): 537-55, and D. Kerr, "African Broadcasting Pioneers and the Origins of Radio Drama at Central African Broadcasting Services," *Critical Arts* 9 (1995): 30-56.

11. Based in part on co-author Oppenheim's fact-finding experiences during three months in Central America and nine months in East Aftrica.

12. Smith, 110.

13. Ibid.

14. Ibid., 27.

15. Ibid., 13, 57, 73-4, and 142.

16. Newsweek, 128 (September 16, 1996), 40; *Cincinnati Enquirer*, (September 15, 1996), A-2.

17. Robert Bellah et al., *Habits of the Heart*, (San Francisco: Harper & Row, 1985), vii, 333 and *passim.*

18. D. R. Anderson et al., "Stressful Life Events and Television Viewing," *Communication Research* 23 (June 1996): 243-60.

19. T. M. MacBeth, "Indirect Effects of Television: Creativity, Persistence, School Achievement, and Participation in other Activities," and R. W. Kubey, "Television Dependence, Diagnosis and Prevention," both in *Tuning in to Young Viewers,* ed. T. M. MacBeth, (Thousand Oaks, C.A.: Sage, 1996), 149-219, 221-60.

20. W. Douglas and B.M. Olson, "Subversion of the American Family? An Examination of Children and Parents in Television Families," *Communication Research* 23 (February 1996): 73-99, and J. Storek and L. Sproull, "As through a glass Darkly: What do People Learn in Videoconferences?" *Human Communications Research* 22 (December 1995): 197-219.

21. *BWSS*, 120-2.

22. Gergen, 224.

23. According to n..17 above, see the pertinent chapters in *Tuning in to ung Viewers*, ed T. M. MacBeth, esp., "Television and Socialization of Young Children," "Indirect Effects of Television," and "Television Dependence, Diagnosis and Prevention."

24. Postman, 137.

25. *BWSS*, 54-7. Also, while vindicating at least a modicum of causal influence of TV viewing upon aggressive behavior, E. F. Dubow and L. S. Miller, in the fifth chapter of *Tuning in to Young Viewers,* rely on the criticized researches of Huesman and Berkowitz. The limitations of these researches have been pointed out by D .Gauntlett, *Moving Experiences*, 33-6. For the link between TV, gender stereotypes and irresponsible sex , see R.W. Kubey's essay in *Tuning in to Young Viewers* and E. R. Gold and R. Speicher, "Marilyn Quayle Meets the Press: Marilyn Loses," *Southern Communication Journal* 61 (Winter 1995): 93-103.

26. *TV Guide* editor, Neil Hickey, "How Much Violence is There?" *TV Guide* vol. 40, # 34 (August 22, 1992): 10-1.

27. David S. Barry, "Growing Up Violent," *Media & Values* 62 (1993): 8-11.

28. Barry L Sherman and Joseph R. Dominick, "Violence and Sex in Music Videos: TV and Rock-n-Roll," *Journal of Communication* 36 (Winter 1986): 79-93.

29. Gauntlett, 112.

30. B. Gunter, *Television and Gender Representation*, Acamedia Research Monographs #14, (London: John Libbey, 1995).

31. K. E. Riggs, "Television Use in a Retirement Community," *Journal of Communications* 46 (Winter 1996): 144-156.

32. J. W. J. Beentijes et al., "Combining Background Media with Doing Homework," *Communication Education* 45 (January 1996): 59-72.

33. A. C. Huston and J. C. Wright, "Television and Socialization in Young Children" in *Tuning in on Young Viewers*, 37-60.

34. R. W. Kubey's article as in n19 above.

35. Scott Peck, *People of the Lie: Hope for Healing Human Evil* (New York: Simon and Schuster, 1983).

36. Josiah Royce, *The Problem of Christianity*, 1 vol. edit. (Chicago: University of Chicago Press, 1968), 348-57.

37. Bellah, et al., vii, 334, and *passim.*

38. Ibid.

39. Dominick, 572.

Chapter 9

ELECTRONIC TECHNOLOGY AND CORPORATE POWER

1. Alice Patricia White, *The Dominant Firm* (Ann Arbor, Michigan: UMI Research Press, 1983).

2. Bagdikian, 4th edit., (1992), 18, 28-30.

3. Dean E. Alger, *The Media and Politics* (Englewood Cliffs, N J.: Prentice Hall, 1989), 75.

4. Bagdikian, 239.

5. Ronald S. Burt, *Corporate Profits and Cooptions* (New York: Academic Press, Inc., 1983), 21.

6. Alger, 90.

7. Burt, 20.

8. Alger, 75.

9. Ibid., 77.

10. Bagdikian, 25.

11. Joyce Nelson, 124-5

12. Bagdikian, 24.

13. *Cincinnati Enquirer* (August 13, 1995).

14. Bagdikian, 21-2.

15. Ibid., 240.

16. Frank Rich, "More Mogul Madness " *New York Times* (November 13, 1996).

17. Murdoch vs Turner, *Newsweek,* (December 11, 1995).

18. Kevin Kelly, "Ephemeral Monoply" *New Perspectives Quarterly,* (Fall 1995): 30.

19. Bill Gates, "Wright on NBC/Microsoft Alliance," *Inside Media,* (May 16, 1995).

20. Alger, 77.

21. "Cable's biggest league," *Newsweek,* (August 26, 1985) : 42.

22. Smith, 113.

23. Daniel R. Levine, "Behind the Cable TV Rip-Off," *Reader's Digest,* (June 1992): 53-57.

24. Bagdikian, 239.

25. Smith, 115.

26. Joshua Cooper Ramo, "Finding God on the Web," *Time,* (December 16, 1996): 60.

27. Nicholas Negroponte in *Time* online interview (March 13, 1995).

28. Smith, 124.

29. Hamid Mowlana, *Global Information and World Communication* (New York: Longman, 1986), 199.

30. Smith., 115.

31. Ibid., 124.

32. Ibid., 125.

33. Wilson P. Dizard, Jr., *The Coming Information Age,* 3rd edit., (New York: Longman, 1989), 183.

34. Smith, 133-4.

35. Hamid Mowlana, 199.

36. Smith, 117.

37. Ibid., 126.

38. Dizard Jr., 186.

39. Smith, 127.

40. Ibid., 128.

41. Dizard Jr., 184, citing Schiller.

42. Ibid., 205.

43. Smith, 133-4.

44. Dizard Jr., 199.

45. Ibid., 200.

46. Smith, 128-9.

47. Ibid., 118-9.

Chapter 10

THE MEDIA'S IMPACT ON THE FUTURE

1. M. I. T.'s media futurist, Nicholas Negroponte describes how profound the results of these discoveries and applications will be in his *Being Digital* (New York; Knopf, 1995), esp., in Part III, "Digital Life" and in the Epilogue. See also Walter Ong, S.J., *Orality and Literacy* (London: Methuen, 1982).

2. The ordinary language usage of the terms "impact" and "influence" was described at the start of Chap. 8. A more carefully detailed look at our electronic future can be found in Negroponte's *Being Digital* (see n. 1) and in Charles Van Doren's *A History of Knowledge* (New York: Carol Publishing Group, 1991), pp. 379-383, 410-412.

3. Joseph R. Dominick, *The Dynamics of Mass Communication*, 5th edit., (New York: McGraw-Hill, 1996), 59.

4. Peter McGrath, "The Web: Infotopia or Marketplace?" *Newsweek* (January 27, 1997): 82.

5. *Time* Online Interview (March 15, 1995) and *Cincinnati Enquirer* (December 2, 1996), A6.

6. Dominich, 333.

7. McGrath, 82.

8. Ibid., 84.

9. Dominick, 224.

10. Ibid., 301.

11. Arno Penzias, *Harmony: Business, Technology & Life After Paperwork* (New York: Harper Business, 1995).

12. Philip Elmer-DeWitt shared these concerns and predictions in a *Time* Online Interview of March 28, 1995.

13. *Newsweek* (January 27, 1997), 82.

14. Dominick, 572.

15. Negroponte, *Being Digital*, 148.

16. This and the next two paragraphs are derived largely from *Time* 146 (August 21, 1995), 38-46.

17. Rev. 13:17

18. Helen Cooper, "White House Seeks Copyright Protection for Text, Software Published On-line," *Wall Street Journal* (September 6, 1995).

19. André Coyne, "Who's in Charge Here? The High-Tech Challenge," *World Press Review* 42 (April 1995): 9; and Charles Arthur, "Internet Anarchy in the UK," *New Scientist* 45 (February 24, 1995): 4.

20. Seth Schiesel, *Los Angles Daily News* (December 21, 1996).

21. Alexander G. Higgins, "Global Agreements Reached to Widen Copyright Law," *The News Tribune*, Tacoma, Washington (December 21, 1996).

22.. Nicholas Negroponte, in a *Time* Online Interview of March 13, 1995.

23. Bruce Olcott, "Will They Take Away My Video-phone If I Get Lousy Ratings? A Proposal for a 'Video Common Carrier' Statue in Post-merger Telecommunications," *Columbia Law Review* 94 (June 1994): 1558-1616.

24. *Time* 146 (September 4, 1995).

25. William James, *The Principles of Psychology*, 2 vols., (New York: Henry Holt, 1890), 1: 243.

26. Dominick, 170.

27. Cf. Wilson P. Dizard, Jr., *The Coming Information Age*, 3rd edit. (New York: Longman, 1989), 233.

28. *Time* 146 (August 28, 1995).

29. Quoted with details from *Newsweek* 129 (January 27, 1997), 66-9.

30. The Vatican's "Pastoral Instruction on Social Communications," parag. #3, *Origins* (March 26, 1992), 671.

31. Ibid., 674.

32. Jon Meacham, "Where Have all the Causes Gone?" *Newsweek* 129 (January 27, 1997): 88.

33. Ibid.

34. The address of the Center for Living Democracy and its *American News Service* is RR #1 Black Fox Road, Brattleboro, VT. 05301; ph. (802) 254-1234.

35. *Time* 146 (August 28, 1995).

Chapter 11

FUTURE CONTROL OF THE MEDIA

1. James W. Carey, *Communication as Culture* (Boston: Unwin Hyman, 1989), 18.

2. Ibid., 20.

3. Ibid., 83.

4. Ibid., 84.

5. Hamid Mowlana, 212-3.

6. Ibid., 214.

7. Ibid., 214.

8. Postman, *Amusing Ourselves to Death,* 162-3.

9. Ibid., 163.

10. Ibid.

11. Lester G. Telser, *Advertising* (New York: Macmillan, 1958), 111.

12. Peter Hamil, "Crack and the Box" in *Impact*, 72.

13. Bagdikian, 56.

14. Smith, 176.

15. Mowlana, 220-1.

INDEX